An Introduction to the Laplace Transform and the *z* Transform

An Introduction to the Laplace Transform and the z Transform

A. C. Grove
Senior Lecturer in Mathematics, Nottingham Polytechnic

Prentice Hall
New York · London · Toronto · Sydney · Tokyo · Singapore

First published 1991 by
Prentice Hall International (UK) Ltd
66 Wood Lane End, Hemel Hempstead
Hertfordshire HP2 4RG
A division of
Simon & Schuster International Group

Typeset in 10 on 12 point Times
by MCS Ltd, Salisbury, Wiltshire, England

Printed and bound in Great Britain
by BPCC Wheatons Ltd, Exeter

Library of Congress Cataloging-in-Publication Data

Grove, A. C. (Anthony C.), 1937–
    An introduction to the Laplace transform and the z transform / by
A.C. Grove.
        p.   cm.
    Includes bibliographical references and index.
    ISBN 0-13-488933-9
    1. Laplace transformation.   2. Z transformation.   I. Title.
QA432.G76   1991
515'.723—dc20   90-22912
                                                              CIP

British Library Cataloguing in Publication Data

Grove, A. C. (Anthony C.) 1937–
    An introduction to the Laplace transform and
the z transform.
    1. Mathematics. Transformations
    I. Title
    515.723

ISBN 0-13-488933-9

---

1 2 3 4 5   95 94 93 92 91

QA432
G76
1991

Contents

v

Preface

This book has arisen from a course of lectures given to engineering students and is intended to assist students in the use of the Laplace transform and the z transform. The background mathematical theory is kept to a minimum.

Laplace transforms and z transforms are convenient methods for solving differential and difference equations respectively; both are amenable to computer implementation. Some indication is given of possible origins of these equations but the main purpose is to discuss methods of solution. However, readers who wish for a deeper mathematical approach should consult one of the more extensive texts.

I would like to thank my colleagues Mr P. W. Moore and Dr A. Sackfield for their helpful criticism and encouragement, Prentice Hall for their sympathetic cooperation, Mrs A. Fullerton and Mrs S. Mohamedali for conscientiously typing the original manuscript and my family May, Allison and David for all the support and interest.

1

The Laplace transform and the inverse Laplace transform

Introduction

One of the most efficient methods for solving certain ordinary and partial differential equations is the use of Laplace transforms. The effectiveness of the transform is its ability to convert many differential equations into algebraic equations. Although other methods provide solutions of equations where the input or forcing function is an exponential and/or sine function the Laplace transform is able to deal with other inputs such as pulses, square waves, point loads, etc., in exactly the same way. The differential equations that arise in the analysis of control systems are particularly suitable for solution by this means.

The Laplace transform is defined in order to develop the necessary techniques but a more detailed introduction is given in Appendix 1.

DEFINITION   Multiply a given function of time $f(t)$ by e^{-st} (s is a parameter) and integrate the product between zero and infinity. The result, if it exists, is denoted by $\mathscr{L}\{f(t)\} = F(s)$ and is called the Laplace transform of $f(t)$, i.e.

$$F(s) = \mathscr{L}\{f(t)\} = \int_0^\infty e^{-st} f(t)\, \mathrm{d}t$$

Some examples are as follows.

1. $f(t) = 1$ $\mathscr{L}\{f(t)\} = \int_0^\infty e^{-st}\, \mathrm{d}t = \left[\frac{e^{-st}}{-s}\right]_0^\infty = \frac{1}{s}$

provided that $s > 0$ so that $e^{-st} \to 0$ as $t \to \infty$.

2. $\mathcal{L}\{e^{-at}\} = \int_0^\infty e^{-st} e^{-at} \, dt = \int_0^\infty e^{-(s+a)t} \, dt = \frac{1}{s+a}$

provided that $s + a > 0$.

3. $\mathcal{L}\{t\} = \int_0^\infty t e^{-st} \, dt$

$$= \left[-\frac{t}{s} e^{-st} \right]_0^\infty + \int_0^\infty \frac{1}{s} e^{-st} \, dt$$

using integration by parts. Hence

$$\mathcal{L}\{t\} = \left[-\frac{1}{s^2} e^{-st} \right]_0^\infty = \frac{1}{s^2}$$

provided that $s > 0$.

4. $\mathcal{L}\{\sin(at)\} = \int_0^\infty e^{-st} \sin(at) \, dt$

$$= \left[\frac{e^{-st}}{s^2 + a^2} \{ -s \sin(at) - a \cos(at) \} \right]_0^\infty$$

$$= \frac{a}{s^2 + a^2}$$

These transforms, and others, constitute a table of transforms which normally consist of a combination of general theorems and the transforms of specific functions. Some of the more common results are given in Table 1.1 but a more comprehensive table is presented in Appendix 2. Note that, since the Laplace transform is defined as an *integral*, the normal rules of integration apply; also the Laplace transform of a function may not exist.

Thus, if K is a constant,

$$\mathcal{L}\{Kf(t)\} = \int_0^\infty K e^{-st} f(t) \, dt$$

$$= K \int_0^\infty e^{-st} f(t) \, dt$$

$$= K \mathcal{L}\{f(t)\}$$

$$= K F(s)$$

Table 1.1 Some Laplace transform pairs

$f(t)$	$\mathcal{L}\{f(t)\} = \displaystyle\int_0^\infty \mathrm{e}^{-st} f(t)\,\mathrm{d}t = F(s)$
K	K/s
t	$1/s^2$
t^n	$\dfrac{n!}{s^{n+1}}$
e^{-at}	$\dfrac{1}{s+a}$
$\sin(\omega t)$	$\dfrac{\omega}{s^2+\omega^2}$
$\cos(\omega t)$	$\dfrac{s}{s^2+\omega^2}$
$\sinh(\omega t)$	$\dfrac{\omega}{s^2-\omega^2}$
$\cosh(\omega t)$	$\dfrac{s}{s^2-\omega^2}$
$Kf(t)$	$KF(s)$
$f(t)+g(t)$	$F(s)+G(s)$

Also

$$\mathcal{L}\{f(t)+g(t)\} = \mathcal{L}\{f(t)\} + \mathcal{L}\{g(t)\}$$

which is the linearity property, but

$$\mathcal{L}\{f(t)g(t)\} \neq \mathcal{L}\{f(t)\}\mathcal{L}\{g(t)\}$$

Functions such as t^{-1} and $\tan(t)$ do not have Laplace transforms since the integrals diverge.

The following example illustrates the use of Table 1.1.

EXAMPLE 1.1

Determine the Laplace transform of

(a) $t^2 + 2t + 3$
(b) $4\sin(3t) + \cosh(5t)$
(c) $\cos^2(t)$

▶

(a) $\quad \mathcal{L}\{t^2 + 2t + 3\} = \mathcal{L}\{t^2\} + 2\mathcal{L}\{t\} + \mathcal{L}\{3\}$

$$= \frac{2}{s^3} + 2\frac{1}{s^2} + \frac{3}{s}$$

(b) $\quad \mathcal{L}\{4 \sin(3t) + \cosh(5t)\} = 4\mathcal{L}\{\sin(3t)\} + \mathcal{L}\{\cosh(5t)\}$

$$= 4\frac{3}{s^2 + 9} + \frac{s}{s^2 - 25}$$

$$= \frac{12}{s^2 + 9} + \frac{s}{s^2 - 25}$$

(c) $\quad \mathcal{L}\{\cos^2(t)\} = \mathcal{L}\{\tfrac{1}{2}[1 + \cos(2t)]\}$

$$= \mathcal{L}\{\tfrac{1}{2}\} + \mathcal{L}\{\tfrac{1}{2}\cos(2t)\}$$

$$= \frac{1}{2s} + \frac{s}{2(s^2 + 4)}$$

EXERCISE 1.1

Use standard formulae to derive the Laplace transform of the following functions:

1	$5t - 2$	**2**	$a + bt + ct^2$
3	$t^3 + 8e^{-t} + 1$	**4**	$2t^2 - 6t$
5	$2 + 4e^{3t}$	**6**	$e^{-at} + e^{-bt}$
7	$a \sin(at) + b \sin(bt)$	**8**	$\cos(at) + \cos(bt)$
9	$5t^2 + 4 \cos(3t)$	**10**	$\sinh(3t)$
11	$\cosh(3t)$	**12**	$t \cos(2t) - \cosh(4t)$
13	$e^{-3t} \cos(2t)$	**14**	$t^4 e^{3t} - e^{-2t} \sin(t)$
15	$(e^t - e^{-t})^2$ and hence $\sinh^2(t)$	**16**	$e^{2t} \cosh(t)$
17	$\cos(at - \alpha)$	**18**	$\sin(bt + \alpha)$
19	$\sin(t)\cos(3t)$	**20**	$\cos(2t) \cos(t)$
21	$\sin^2(t)$		

where a, b, c and α are constant. In questions 19, 20 and 21 use a suitable trigonometric identity.

The inverse Laplace transform

In practice it is important to be able to recover $f(t)$ from its Laplace transform. It can be shown that $f(t)$ is uniquely determined by $F(s)$. If $F(s) = \mathcal{L}\{f(t)\}$ then $f(t) = \mathcal{L}^{-1}\{F(s)\}$ where $\mathcal{L}^{-1}$ denotes the inverse Laplace transform. For example

$$\mathcal{L}^{-1}\left\{\frac{1}{s^2 + 4}\right\} = \tfrac{1}{2}\sin(2t)$$

Rarely do the functions of s appear in standard form. In general $F(s)$ is a ratio of two polynomials in s which can be expanded in partial fractions provided that the degree of the numerator is less than the degree of the denominator (a proper rational function).

EXAMPLE 1.2

Determine the inverse Laplace transform of

(a) $\dfrac{2s + 3}{s^2 + 4s + 13}$ (b) $\dfrac{s - 6}{(s - 1)(s - 2)}$

(c) $\dfrac{4s - 9}{(s - 1)^2(s - 2)}$

(a) $\mathcal{L}^{-1}\left\{\dfrac{2s + 3}{s^2 + 4s + 13}\right\} = \mathcal{L}^{-1}\left\{\dfrac{2s + 3}{(s + 2)^2 + 9}\right\}$

on completing the square. Then

$$\mathcal{L}^{-1}\left\{\frac{2s + 3}{(s + 2)^2 + 9}\right\} = \mathcal{L}^{-1}\left\{\frac{2(s + 2)}{(s + 2)^2 + 9}\right\} - \mathcal{L}^{-1}\left\{\frac{1}{(s + 2)^2 + 9}\right\}$$

$$= 2e^{-2t}\cos(3t) - \tfrac{1}{3}e^{-2t}\sin(3t)$$

(b) $\mathcal{L}^{-1}\left\{\dfrac{s - 6}{(s - 1)(s - 2)}\right\} = \mathcal{L}^{-1}\left\{\dfrac{5}{s - 1} - \dfrac{4}{s - 2}\right\}$

by partial fractions. Thus

$$\mathcal{L}^{-1}\left\{\frac{s - 6}{(s - 1)(s - 2)}\right\} = 5e^t - 4e^{2t}$$

(c) $\mathcal{L}^{-1}\left\{\dfrac{4s - 9}{(s - 1)^2(s - 2)}\right\} = \mathcal{L}^{-1}\left\{\dfrac{1}{s - 1} + \dfrac{5}{(s - 1)^2} - \dfrac{1}{s - 2}\right\}$

by partial fractions. Thus

$$\mathcal{L}^{-1}\left\{\frac{4s - 9}{(s - 1)^2(s - 2)}\right\} = e^t + 5te^t - e^{2t}$$

EXERCISE 1.2

Determine the inverse Laplace transform of each of the following functions of s:

1 $\dfrac{1}{s^4}$

2 $\dfrac{s+1}{s^3}$

3 $\dfrac{2}{s^2} - \dfrac{3}{s} + \dfrac{5}{s+1}$

4 $\dfrac{1}{2s+1}$

5 $\dfrac{1}{3s+4}$

6 $\dfrac{s}{4s^2+1}$

7 $\dfrac{1}{s^2-4}$

8 $\dfrac{9s}{s^2-16}$

9 $\dfrac{2s-5}{s^2+9}$

10 $\dfrac{s+3}{s^2+1}$

11 $\dfrac{3}{(s+3)(s-2)}$

12 $\dfrac{1}{(s+3)^2+3^2}$

13 $\dfrac{s}{(s^2+9)^2}$

14 $\dfrac{5}{(s^2+16)^2}$

15 $\dfrac{7!}{(s-3)^8}$

16 $\dfrac{s}{s^2+2s-3}$

17 $\dfrac{s}{s^2+2s-2}$

18 $\dfrac{1}{(s^2+6s+13)^2}$

19 $\dfrac{3s-4}{(s^2+9)(s-2)}$

20 $\dfrac{7s^2+2s}{(s^2+4)(s^2-9)}$

21 $\dfrac{1}{s^2(s^2+1)(s^2+4)}$

2

The transforms of derivatives and the application to differential equations

Introduction

In the process of formulating practical problems by mathematical equations, i.e. mathematical modeling, many kinds of differential equations can arise. Examination of such equations has led to the conclusion that there are certain definite methods by which many of them can be solved. It is not the intention to provide detailed information on mathematical modeling here but to emphasize the similarity between many different applications and to discuss a method of solving some of the equations. A number of these applications will be used in subsequent sections.

In many instances the mathematical model is obtained from scientific laws. Many of the general laws of nature find their most natural expression in the language of differential equations. These equations have to be solved to give expressions relating the variables involved without the use of derivatives. It is important, however, to realize that in obtaining the mathematical model it may be necessary to make certain assumptions. For example, it is sometimes assumed that resistance to motion is proportional to velocity – a valid assumption under certain conditions. The derived equation and its solution must therefore be used under the same conditions. It is also important to note that different statements (laws) give rise to similar types of differential equations. The equations of some typical examples are given on p8.

Electrical circuits

The application of Kirchhoff's law to an LCR circuit yields the differential equations

$$L\frac{di}{dt} + Ri + \frac{q}{C} = E \qquad i = \frac{dq}{dt}$$

which on elimination of i gives

$$L\frac{d^2q}{dt^2} + R\frac{dq}{dt} + \frac{q}{C} = E$$

or on elimination of q gives

$$L\frac{d^2i}{dt^2} + R\frac{di}{dt} + \frac{i}{C} = \frac{dE}{dt}$$

Circuits with several loops generate simultaneous differential equations

$$L\frac{di_1}{dt} + Ri_3 = E$$

$$\frac{1}{C}i_2 - R\frac{di_3}{dt} = 0$$

$$i_1 = i_2 + i_3$$

Particle dynamics

If a particle of mass m is suspended from a fixed point by a spring of constant k then the equation of motion is

$$m\frac{d^2x}{dt^2} = -kx$$

where x is the displacement from the equilibrium position.

This equation is derived by the application of Newton's second law assuming that the force exerted by the spring is proportional to the extension (Hooke's law). The inclusion of damping, assumed to be proportional to the velocity, and external forces generate further terms in the equation. Similarly the motion of several connected particles can be described by a set of simultaneous equations.

The motion of a particle projected from the surface of the earth can be described, relative to suitable three-demensional axes, by the equations

$$\frac{d^2x}{dt^2} = 2\omega \sin(\theta) \frac{dy}{dt}$$

$$\frac{d^2y}{dt^2} = -2\omega \sin(\theta) \frac{dx}{dt} - 2\omega \cos(\theta) \frac{dz}{dt}$$

$$\frac{d^2z}{dt^2} = 2\omega \cos(\theta) \frac{dy}{dt} - g$$

Servomechanisms

The error between a required angle ϕ and an actual angle θ in a radar control system is

$$\varepsilon = \theta - \phi$$

If the correcting torque is assumed to be proportional to the error then the differential equation is of the form

$$\frac{d^2\theta}{dt^2} = -k(\theta - \phi)$$

The negative sign is required because the torque must oppose the error.

Biology and chemistry

Differential equations describe, for example, a mixture of chemicals, radioactive decay, population growth, epidemics, predator–prey relationships and Newton's law of cooling. Typical equations are as follows:

$$\frac{dT}{dt} = -K(T - T_A)$$

$$\frac{d^2\theta}{dt^2} = K(\theta - \theta_A)$$

$$\frac{dx}{dt} = 3x - 2y \qquad \frac{dy}{dt} = -2x + 3y$$

Structures

The bending of a beam can be modeled by the use of one of the following equations, depending on the loading.

$$EI \frac{d^2 y}{dx^2} = M \qquad EI \frac{d^4 y}{dx^4} = w$$

$$EI \frac{d^2 y}{dx^2} + Py = 0 \qquad EI \frac{d^4 y}{dx^4} + ky = f(x)$$

Economics

Commodity prices depend on quantities such as supply and inflation factors. The resulting differential equations are of the form

$$\frac{dP}{dt} = F(t) - k_1(S - S_0) \qquad \frac{dS}{dt} = k_2(P - P_0)$$

A linear second-order differential equation with constant coefficients is thus of the form

$$a \frac{d^2 x}{dt^2} + b \frac{dx}{dt} + cx = f(t)$$

The transforms of derivatives

The transforms of the first, second, third and fourth derivatives are given in Table 2.1 and are numbered 2, 3, 4 and 5 respectively in

Table 2.1 Laplace transforms of derivatives

$f(t)$	$F(s)$
$\dfrac{d}{dt}[f(t)]$	$sF(s) - f(0)$
$\dfrac{d^2}{dt^2}[f(t)]$	$s^2 F(s) - sf(0) - f'(0)$
$\dfrac{d^3}{dt^3}[f(t)]$	$s^3 F(s) - s^2 f(0) - sf'(0) - f''(0)$
$\dfrac{d^4}{dt^4}[f(t)]$	$s^4 F(s) - s^3 f(0) - s^2 f'(0) - sf''(0) - f'''(0)$

Appendix 2. Note that $F(s) = \mathscr{L}\{f(t)\}$ as usual and that $f(0), f'(0), f''(0), \ldots$ are the numerical values of $f(t), f'(t), f''(t), \ldots$ when $t = 0$, where a prime denotes differentiation with respect to t. These results are obtained from the definition of the Laplace transform and applying the rule for integration by parts. For example

$$\mathscr{L}\left\{\frac{\mathrm{d}}{\mathrm{d}t}\ [f(t)]\right\} = \int_0^\infty \mathrm{e}^{-st}\ \frac{\mathrm{d}}{\mathrm{d}t}\ [f(t)]\ \mathrm{d}t$$

$$= \left[\mathrm{e}^{-st}f(t)\right]_0^\infty - \int_0^\infty - s\ \mathrm{e}^{-st}f(t)\ \mathrm{d}t$$

$$= -f(0) + s\int_0^\infty \mathrm{e}^{-st}f(t)\ \mathrm{d}t$$

provided that $\mathrm{e}^{-st}f(t) \to 0$ as $t \to \infty$. Thus

$$\mathscr{L}\left\{\frac{\mathrm{d}}{\mathrm{d}t}\ [f(t)]\right\} = -f(0) + s\mathscr{L}\{f(t)\}$$

$$= sF(s) - f(0)$$

Similarly

$$\mathscr{L}\left\{\frac{\mathrm{d}^2}{\mathrm{d}t^2}\ [f(t)]\right\} = \int_0^\infty \mathrm{e}^{-st}\ \frac{\mathrm{d}^2}{\mathrm{d}t^2}\ [f(t)]\ \mathrm{d}t$$

$$= \left[\mathrm{e}^{-st}f'(t)\right]_0^\infty + s\int_0^\infty \mathrm{e}^{-st}\ \frac{\mathrm{d}}{\mathrm{d}t}\ [f(t)]\ \mathrm{d}t$$

$$= -f'(0) + s\mathscr{L}\left\{\frac{\mathrm{d}}{\mathrm{d}t}\ [f(t)]\right\}$$

provided that $\mathrm{e}^{-st}f'(t) \to 0$ as $t \to \infty$. Hence

$$\mathscr{L}\left\{\frac{\mathrm{d}^2}{\mathrm{d}t^2}\ [f(t)]\right\} = -f'(0) + s[s\mathscr{L}\{f(t)\} - f(0)]$$

$$= s^2\mathscr{L}\{f(t)\} - sf(0) - f'(0)$$

$$= s^2F(s) - sf(0) - f'(0)$$

If $f(t)$ is a known function of t then the results may be used to yield further standard formulae. In the solution of a differential equation such as

$$\frac{\mathrm{d}^2x}{\mathrm{d}t^2} + 3\ \frac{\mathrm{d}x}{\mathrm{d}t} + 5x = \sin(2t)$$

x is an unknown function of t which we are attempting to find. Its Laplace transform is therefore an unknown function of s, denoted by $\bar{x}$ or $X(s)$. Hence

Function	Laplace transform
x	$\bar{x}$ or $X(s)$
$\dfrac{dx}{dt}$	$s\bar{x} - x_0$
$\dfrac{d^2x}{dt^2}$	$s^2\bar{x} - sx_0 - x_0'$

where x_0 and x_0' denote the initial values of x and dx/dt respectively. These expressions are algebraic expressions in s; the transformed equation will thus be an algebraic equation in s which can be solved for $\bar{x}$. The solution x in terms of t is then obtained by inversion.

Method

1. Transform the differential equation using all the necessary rules and standard forms.
2. Solve the resulting equation (usually algebraic) for the unknown as a function of s.
3. Invert the function of s to obtain the required solution.

EXAMPLE 2.1

Solve the equation

$$\frac{d^2x}{dt^2} + 3\frac{dx}{dt} + 2x = 0$$

given that $x = 4$, $dx/dt = -3$ when $t = 0$.

Transforming the equation

$$\mathscr{L}\left\{\frac{d^2x}{dt^2}\right\} + 3\mathscr{L}\left\{\frac{dx}{dt}\right\} + 2\mathscr{L}\{x\} = \mathscr{L}\{0\}$$

$$(s^2\bar{x} - sx_0 - x_0') + 3(s\bar{x} - x_0) + 2\bar{x} = 0$$

and since $x_0 = 4$, $x_0' = -3$ the equation can be solved for $\bar{x}$

$$\bar{x}(s^2 + 3s + 2) = 4s + 9$$

$$\bar{x} = \frac{4s + 9}{(s + 2)(s + 1)} = \frac{5}{(s + 1)} - \frac{1}{(s + 2)}$$

by partial fractions. On inversion this gives

$$x = 5e^{-t} - e^{-2t}$$

EXAMPLE 2.2

Solve the equation

$$\frac{d^2x}{dt^2} + 3\frac{dx}{dt} + 2x = e^{-2t}$$

Since no *initial* conditions are given it is necessary to retain x_0 and x_0' as unknown constants. Transforming the equation

$$(s^2\bar{x} - sx_0 - x_0') + 3(s\bar{x} - x_0) + 2\bar{x} = \frac{1}{s+2}$$

$$\bar{x} = \frac{1}{(s+2)(s^2+3s+2)} + \frac{sx_0 + x_0' + 3x_0}{s^2+3s+2}$$

$$= \frac{1 + x_0s(s+2) + (x_0' + 3x_0)(s+2)}{(s+1)(s+2)^2}$$

since $s^2 + 3s + 2 = (s+1)(s+2)$. In partial fractions this is

$$\bar{x} = \frac{A}{s+1} + \frac{B}{s+2} + \frac{C}{(s+2)^2}$$

from which

$$A = 1 + 2x_0 + x_0'$$

$$B = -x_0' - x_0 - 1$$

$$C = -1$$

Inverting,

$$x = Ae^{-t} + Be^{-2t} - te^{-2t}$$

If non-initial conditions are known such as $x = 1$, $dx/dt = 3$ when $t = 4$, then the constants A and B can be determined by substitution.

Simultaneous differential equations

The solution of a set of simultaneous differential equations is achieved using the same procedure. The transformed equations are again normally algebraic and can be solved by elimination or even matrix methods.

EXAMPLE 2.3

Solve the simultaneous equations

$$\frac{dx}{dt} + 3x + 5y = e^{-t} \tag{2.1}$$

$$\frac{dy}{dt} - 3y - 5x = 0 \tag{2.2}$$

given that $x = 0$, $y = 0$ when $t = 0$.

The transformed equations are

$$s\bar{x} + 3\bar{x} + 5\bar{y} = \frac{1}{s+1}$$

$$s\bar{y} - 3\bar{y} - 5\bar{x} = 0$$

$$\bar{x}(s+3) + 5\bar{y} = \frac{1}{s+1} \tag{2.3}$$

$$-5\bar{x} + (s-3)\bar{y} = 0 \tag{2.4}$$

Eliminating $\bar{x}$ by the operation $5 \times$ equation (2.3) + $(s+3) \times$ equation (2.4) gives

$$\bar{y} = \frac{5}{(s+1)(s^2+16)} = \frac{5}{17(s+1)} + \frac{-5s+5}{17(s^2+16)}$$

which on inversion gives

$$y = \tfrac{5}{17} e^{-t} + \tfrac{5}{68} \sin(4t) - \tfrac{5}{17} \cos(4t)$$

Equations (2.3) and (2.4) can now be solved for $\bar{x}$ and a further inversion yields an expression for x in terms of t.

   A more convenient method is to note that the second of the *original* differential equations gives

$$5x = \frac{dy}{dt} - 3y \tag{2.2}$$

which on substitution for y and dy/dt from above gives

$$5x = -\tfrac{20}{17} e^{-t} + \tfrac{20}{17} \cos(4t) + \tfrac{65}{68} \sin(4t)$$

$$x = -\tfrac{4}{17} e^{-t} + \tfrac{4}{17} \cos(4t) + \tfrac{13}{68} \sin(4t)$$

This method of substituting into one of the original differential equations in order to avoid the necessity of inverting two separate functions is possible provided that *one* of the equations contains only one derivative, not two (i.e. either dx/dt or dy/dt but not both).

EXAMPLE 2.4

Solve the simultaneous differential equations

$$\frac{dx}{dt} + 2\frac{dy}{dt} + 3x = e^{-t} \tag{2.5}$$

$$3\frac{\mathrm{d}x}{\mathrm{d}t} + \frac{\mathrm{d}y}{\mathrm{d}t} + 3y + x = 0 \tag{2.6}$$

given that $x = 0$, $y = 0$ when $t = 0$.

Note that both equations contain $\mathrm{d}x/\mathrm{d}t$ and $\mathrm{d}y/\mathrm{d}t$. In this case generate a third differential equation by eliminating either $\mathrm{d}x/\mathrm{d}t$ or $\mathrm{d}y/\mathrm{d}t$ (free choice). For example, eliminating $\mathrm{d}y/\mathrm{d}t$ by multiplying equation (2.6) by 2 and subtracting equation (2.5) gives

$$5\frac{\mathrm{d}x}{\mathrm{d}t} + 6y - x = -\mathrm{e}^{-t} \tag{2.7}$$

which will give y directly provided that x can be found.

Returning to the original differential equations (2.5) and (2.6) and transforming we obtain

$$(s\bar{x} - x_0) + 2(s\bar{y} - y_0) + 3\bar{x} = \frac{1}{s+1}$$

$$3(s\bar{x} - x_0) + (s\bar{y} - y_0) + 3\bar{y} + \bar{x} = 0$$

Rearranging we have

$$(s+3)\bar{x} + 2s\bar{y} = \frac{1}{s+1} \tag{2.8}$$

$$(3s+1)\bar{x} + (s+3)\bar{y} = 0 \tag{2.9}$$

Now eliminate $\bar{y}$ (we need $\bar{x}$) by multiplying equation (2.8) by $s+3$ and subtract equation (2.9) multiplied by $2s$:

$$\bar{x}[(s+3)^2 - 2s(3s+1)] = \frac{s+3}{s+1}$$

$$\bar{x} = \frac{s+3}{(s+1)(-5s^2 + 4s + 9)}$$

$$= \frac{-(s+3)}{(s+1)(5s-9)(s+1)}$$

$$= -\frac{30}{49(5s-9)} + \frac{6}{49(s+1)} + \frac{1}{7(s+1)^2}$$

Inverting, we obtain

$$x = -\tfrac{6}{49}\,\mathrm{e}^{1.8t} + \tfrac{6}{49}\,\mathrm{e}^{-t} + \tfrac{1}{7}\,t\mathrm{e}^{-t}$$

This result can then be used to obtain y by substitution into equation (2.7)

$$6y = x - \mathrm{e}^{-t} - 5\frac{\mathrm{d}x}{\mathrm{d}t}$$

$$= \tfrac{48}{49}\,\mathrm{e}^{1.8t} - \tfrac{48}{49}\,\mathrm{e}^{-t} + \tfrac{6}{7}\,t\mathrm{e}^{-t}$$

$$y = \tfrac{8}{49}\,\mathrm{e}^{1.8t} - \tfrac{8}{49}\,\mathrm{e}^{-t} + \tfrac{1}{7}\,t\mathrm{e}^{-t}$$

EXAMPLE 2.5

If an electron is projected into a uniform magnetic field perpendicular to its direction of motion its path is given by

$$m\frac{d^2y}{dt^2} = -\frac{He}{c}\frac{dx}{dt} \qquad m\frac{d^2x}{dt^2} = \frac{He}{c}\frac{dy}{dt}$$

where m, c, H and e are constants. If $x = 0$, $dx/dt = u$, $y = 0$ and $dy/dt = 0$ at $t = 0$, determine x and y in terms of t.

The transformed equations are

$$m(s^2\bar{y} - sy_0 - y_0') = -\frac{He}{c}(s\bar{x} - x_0) \tag{2.10}$$

$$m(s^2\bar{x} - sx_0 - x_0') = \frac{He}{c}(s\bar{y} - y_0) \tag{2.11}$$

Inserting the initial conditions and rearranging gives

$$s^2\bar{y} + \frac{He}{cm}s\bar{x} = 0 \tag{2.12}$$

$$\frac{He}{cm}s\bar{y} - s^2\bar{x} = -u \tag{2.13}$$

Now eliminate $\bar{y}$

$$\frac{He}{cm}s\bar{y} + \left(\frac{He}{cm}\right)^2\bar{x} = 0 \tag{2.14}$$

$$\frac{He}{cm}s\bar{y} - s^2\bar{x} = -u \tag{2.15}$$

$$\bar{x} = \frac{u}{s^2 + (He/cm)^2}$$

Therefore

$$x = \frac{ucm}{He}\sin\left(\frac{He}{cm}t\right)$$

Similarly

$$y = -\frac{ucm}{He}\left[1 - \cos\left(\frac{He}{cm}t\right)\right]$$

EXERCISE 2

Solve the following equations:

1 $\dfrac{dy}{dt} + 4y = 1$

given that $y = 2$ when $t = 0$.

2 $\dfrac{d^2y}{dt^2} + 5\dfrac{dy}{dt} + 6y = 3$

given that $y = 2$, $dy/dt = 0$ when $t = 0$.

3 $\dfrac{d^2y}{dt^2} + 2\dfrac{dy}{dt} + y = \sin(t)$

given that $y = 3$, $dy/dt = 1$ when $t = 0$.

4 $\dfrac{dT}{d\theta} - \mu T = 0$

given that $T = T_0$ when $\theta = 0$ and μ and T_0 are constant.

5 $\dfrac{d^2r}{d\theta^2} + 4\dfrac{dr}{d\theta} + 4r = 0$

given that $r = 1$, $dr/d\theta = 2$ when $\theta = 0$.

6 $\dfrac{d^2x}{dt^2} + 16x = 32$

given that $x = 3$, $dx/dt = 2$ when $t = 0$.

7 $\dfrac{d^2x}{dt^2} + 5\dfrac{dx}{dt} + 4x = e^{3t}$

given that $x = 0$, $dx/dt = -2$ when $t = 0$.

8 $\dfrac{dx}{dt} + 2x + \dfrac{dy}{dt} = 0 \qquad \dfrac{dx}{dt} + 2\dfrac{dy}{dt} + 3y = e^{-t}$

given that $x = -1$, $y = 2$ when $t = 0$.

9 $\dfrac{dx}{dt} + \dfrac{dy}{dt} + x + y = 3\sin(2t) \qquad \dfrac{dy}{dt} + 5x + 4y = 0$

given that $x = 0$, $y = 0$ when $t = 0$.

10 $\dfrac{dx}{dt} + x - 2y = 3 \qquad \dfrac{dy}{dt} - 2x + y = e^{2t}$

given that $x = 2$, $y = -1$ when $t = 0$.

11 The displacements x and y of two connected particles are given by the equations

$$\frac{d^2x}{dt^2} - 3x - 4y = 0$$

$$\frac{d^2y}{dt^2} + x + y = 0$$

Determine the general solution for x and y in terms of t.

12 The currents i_1 and i_2 in connected loops satisfy the differential equations

$$2\frac{di_1}{dt} - \frac{di_2}{dt} + i_1 = 20\cos(t)$$

$$2\frac{di_2}{dt} - \frac{di_1}{dt} + i_2 = 0$$

Determine each current in terms of t if initially both are zero.

13 A light rod is clamped horizontally at one end ($x = 0$), is freely hinged at the other ($x = l$) and is subject to a horizontal thrust P at $x = 0$. If G is the couple applied at the clamped end the deflection y satisfies the differential equation

$$EI\frac{d^2y}{dx^2} + Py = G\left(1 - \frac{x}{l}\right)$$

where E, I, P, G and l are constant. Show that

$$y = \frac{G}{P}\left[\frac{\sin(nx) - nx}{nl} + 1 - \cos(nx)\right]$$

where $n^2 = P/EI$, and hence that

$$\tan(nl) = nl$$

3

Some useful theorems

There are a number of general results which facilitate the determination and application of the Laplace transform. Some of the more important ones are discussed here but more comprehensive tables are available (Erdélyi, 1954; Spiegel, 1965) as are formal proofs (Doetsch, 1971; Watson, 1981). Each property is also referenced to the table of transforms (Appendix 2).

In the use of any general results it is absolutely necessary to have a clear understanding of the functional notation $f(t)$, $F(s)$, $F(s+a)$, and of the relationship between $f(t)$ and $F(s)$:

$F(s) = \mathscr{L}\{f(t)\}$ $f(t) = \mathscr{L}^{-1}\{F(s)\}$

$F(s+a)$ is obtained from $F(s)$ on replacing s by $s+a$

$F(s)$ is obtained from $F(s+a)$ on omitting a

Linearity property (1)

If $\mathscr{L}\{f(t)\}$ and $\mathscr{L}\{g(t)\}$ exist, then $\mathscr{L}\{af(t)+bg(t)\}$, where a and b are constants, exists and

$$\mathscr{L}\{af(t)+bg(t)\} = a\mathscr{L}\{f(t)\} + b\mathscr{L}\{g(t)\}$$

since, from the definition,

$$\int_0^\infty e^{-st}[af(t)+bg(t)] \ dt = a\int_0^\infty e^{-st}f(t) \ dt + b\int_0^\infty e^{-st}g(t) \ dt$$

19

First shifting property (exponential multiplier) (7)

If $\mathscr{L}\{f(t)\} = F(s)$ then $\mathscr{L}\{e^{-at}f(t)\} = F(s+a)$. From the definition

$$\mathscr{L}\{e^{-at}f(t)\} = \int_0^\infty e^{-st}e^{-at}f(t)\ dt$$

$$= \int_0^\infty e^{-(s+a)t}f(t)\ dt$$

$$= F(s+a)$$

since $F(s) = \int_0^\infty e^{-st}f(t)\ dt$. Similarly

$$\mathscr{L}^{-1}\{F(s+a)\} = e^{-at}f(t)$$

EXAMPLE 3.1

From $\mathscr{L}\{t^n\} = n!/s^{(n+1)}$

$$\mathscr{L}\{e^{-at}t^n\} = \frac{n!}{(s+a)^{n+1}}$$

EXAMPLE 3.2

Determine

$$\mathscr{L}^{-1}\left\{\frac{1}{(s+2)^4}\right\}$$

If $F(s+a) = 1/(s+2)^4$ then $F(s) = 1/s^4$ and $f(t) = \mathscr{L}^{-1}\{F(s)\}$

$$\mathscr{L}^{-1}\left\{\frac{1}{(s+2)^4}\right\} = e^{-2t}\mathscr{L}^{-1}\left\{\frac{1}{s^4}\right\} = e^{-2t}\frac{t^3}{3!}$$

The transform of an integral (6)

If $F(s) = \mathscr{L}\{f(t)\}$ then

$$\mathscr{L}\left\{\int_0^t f(t)\ dt\right\} = \frac{1}{s}F(s)$$

Similarly

$$\mathscr{L}^{-1}\left\{\frac{1}{s}F(s)\right\} = \int_0^t f(t)\,\mathrm{d}t$$

EXAMPLE 3.3

Determine

$$\mathscr{L}\left\{\int_0^t e^{-t}\sin(3t)\,\mathrm{d}t\right\}$$

$$\mathscr{L}\left\{\int_0^t e^{-t}\sin(3t)\,\mathrm{d}t\right\} = \frac{1}{s}\mathscr{L}\{e^{-t}\sin(3t)\}$$

$$= \frac{1}{s}\frac{3}{(s+1)^2+9} = \frac{3}{s(s^2+2s+10)}$$

Alternatively evaluate the integral and then determine the transform of the resulting function of t.

$$\int_0^t e^{-t}\sin(3t)\,\mathrm{d}t = \frac{1}{10}\left\{3 + e^{-t}[-\sin(3t) - 3\cos(3t)]\right\}$$

$$\mathscr{L}\left\{\int_0^t e^{-t}\sin(3t)\,\mathrm{d}t\right\} = \mathscr{L}\left\{\frac{3}{10} + \frac{e^{-t}}{10}[-\sin(3t) - 3\cos(3t)]\right\}$$

$$= \frac{3}{10s} - \frac{3}{10[(s+1)^2+9]} - \frac{3(s+1)}{10[(s+1)^2+9]}$$

$$= \frac{3(s^2+2s+10) - 3s - 3s(s+1)}{10s(s^2+2s+10)}$$

$$= \frac{3}{s(s^2+2s+10)}$$

EXAMPLE 3.4

Determine

$$\mathscr{L}\left\{\int_0^t i\,\mathrm{d}t\right\}$$

where i is an unspecified function of t.

$$\mathscr{L}\left\{\int_0^t i\,\mathrm{d}t\right\} = \frac{1}{s}\mathscr{L}\{i\} = \frac{1}{s}\bar{\imath}$$

where $\bar{\imath} = \mathscr{L}\{i\}$.

EXAMPLE 3.5

Evaluate

$$\mathscr{L}^{-1}\left\{\frac{\omega}{s[(s-a)^2+\omega^2]}\right\}$$

where a and ω are constants.

Using (6)

$$\mathscr{L}^{-1}\left\{\frac{\omega}{s[(s-a)^2+\omega^2]}\right\} = \int_0^t \mathscr{L}^{-1}\left\{\frac{\omega}{[(s-a)^2+\omega^2]}\right\} \, dt$$

$$= \int_0^t e^{at}\sin(\omega t) \, dt$$

$$= \frac{1}{a^2+\omega^2}\{\omega + e^{at}[a\sin(\omega t) - \omega\cos(\omega t)]\}$$

EXAMPLE 3.6

Determine

$$\mathscr{L}^{-1}\left\{\frac{1}{s(s+a)^3}\right\}$$

Using (6)

$$\mathscr{L}^{-1}\left\{\frac{1}{s(s+a)^3}\right\} = \int_0^t \mathscr{L}^{-1}\left\{\frac{1}{(s+a)^3}\right\} \, dt$$

$$= \frac{1}{2}\int_0^t t^2 e^{-at} \, dt$$

$$= \frac{1}{2}\left(-\frac{t^2}{a}e^{-at} - \frac{2t}{a^2}e^{-at} - \frac{2}{a^3}e^{-at} + \frac{2}{a^3}\right)$$

on integrating by parts twice.

Note that in Examples 3.5 and 3.6 the same results could have been obtained by other methods. For example

$$\frac{1}{s(s+a)^3} \equiv \frac{A}{s} + \frac{B}{s+a} + \frac{C}{(s+a)^2} + \frac{D}{(s+a)^3}$$

in partial fractions. The constants A, B, C and D are evaluated and each term is then inverted.

EXAMPLE 3.7

Solve the differential equation

$$L\frac{di}{dt} + Ri + \frac{1}{C}\int_0^t i\ dt = E$$

for the current i in a circuit if L, C, R and E are constants and $i = 0$ when $t = 0$.

Transforming the equation

$$L(s\bar{i} - i_0) + R\bar{i} + \frac{1}{Cs}\ \bar{i} = \frac{E}{s}$$

$$\bar{i}\left(Ls + R + \frac{1}{Cs}\right) = \frac{E}{s}$$

$$\bar{i} = \frac{E}{(Ls^2 + Rs + 1/C)}$$

$$= E\left\{L\left[\left(s + \frac{R}{2L}\right)^2 + \left(\frac{1}{LC} - \frac{R^2}{4L^2}\right)\right]\right\}^{-1}$$

If we assume that $R^2 < 4L/C$, $\alpha = R/2L$ and $\beta^2 = 1/LC - R^2/4L^2$ then $\beta^2 > 0$ and the solution on inversion becomes

$$i = \frac{E}{L\beta}\ e^{-\alpha t}\ \sin(\beta t)$$

However, if $R^2 > 4L/C$, $\alpha = R/2L$, $\gamma^2 = R^2/4L^2 - 1/LC$ then $\gamma^2 > 0$. Hence

$$\bar{i} = \frac{E}{L[(s + \alpha)^2 - \gamma^2]}$$

and the solution becomes

$$i = \frac{E}{L\gamma}\ e^{-\alpha t}\ \sinh(\gamma t)$$

Differentiation of transforms

If $F(s) = \mathscr{L}\{f(t)\}$ then

$$\mathscr{L}\{-tf(t)\} = \frac{d}{ds}\ \{F(s)\}$$

and more generally

$$\mathscr{L}\{(-t)^n f(t)\} = \frac{d^n}{ds^n}\ \{F(s)\} \qquad n = 1, 2, \ldots$$

Note also therefore that

$$\mathscr{L}^{-1}\{F(s)\} = -\frac{1}{t}\,\mathscr{L}^{-1}\left\{\frac{\mathrm{d}}{\mathrm{d}s}\,[F(s)]\right\}$$

If $(1/t)f(t)$ has a limit as $t \to 0$ from the right, then

$$\mathscr{L}\left\{\frac{1}{t}\,f(t)\right\} = \int_s^\infty F(s)\,\mathrm{d}s$$

or in inverse form

$$\mathscr{L}^{-1}\{F(s)\} = t\,\mathscr{L}^{-1}\left\{\int_s^\infty F(s)\,\mathrm{d}s\right\}$$

EXAMPLE 3.8

Find $\mathscr{L}\{t^2 \sin(2t)\}$.

$$\mathscr{L}\{t^2 \sin(2t)\} = (-1)^2\,\frac{\mathrm{d}^2}{\mathrm{d}s^2}\,[\mathscr{L}\{\sin(2t)\}]$$

$$= (-1)^2\,\frac{\mathrm{d}^2}{\mathrm{d}s^2}\left(\frac{2}{s^2+4}\right)$$

$$= \frac{12s^2 - 16}{(s^2+4)^3}$$

EXAMPLE 3.9

Find y if

$$\bar{y} = \ln\left(\frac{s+1}{s-1}\right)$$

$$y = \mathscr{L}^{-1}\left\{\ln\left(\frac{s+1}{s-1}\right)\right\}$$

$$= -\frac{1}{t}\,\mathscr{L}^{-1}\left\{\frac{\mathrm{d}}{\mathrm{d}s}\left[\ln\left(\frac{s+1}{s-1}\right)\right]\right\}$$

$$= -\frac{1}{t}\,\mathscr{L}^{-1}\left\{\frac{1}{s+1} - \frac{1}{s-1}\right\}$$

$$= -\frac{1}{t}\,(\mathrm{e}^{-t} - \mathrm{e}^t) = \frac{2}{t}\,\sinh(t)$$

EXAMPLE 3.10

Find

$$\mathscr{L}\left\{\frac{1}{t}\sin(kt)\right\}$$

$$\mathscr{L}\left\{\frac{1}{t}\sin(kt)\right\} = \int_s^\infty \mathscr{L}\{\sin(kt)\}\ \mathrm{d}s$$

$$= \int_s^\infty \frac{k}{s^2+k^2}\ \mathrm{d}s$$

$$= \left[\tan^{-1}\left(\frac{s}{k}\right)\right]_s^\infty = \frac{\pi}{2} - \tan^{-1}\left(\frac{s}{k}\right)$$

$$= \cot^{-1}\left(\frac{s}{k}\right)$$

from Figure 3.1.

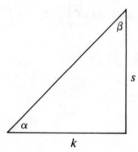

Figure 3.1

EXAMPLE 3.11

Find y if $\bar{y} = s/(s^2-1)^2$.

$$y = t\,\mathscr{L}^{-1}\left\{\int_s^\infty \frac{s}{(s^2-1)^2}\ \mathrm{d}s\right\}$$

$$= t\,\mathscr{L}^{-1}\left\{\left[\frac{-1}{2(s^2-1)}\right]_s^\infty\right\}$$

$$= t\,\mathscr{L}^{-1}\left\{\frac{1}{2(s^2-1)}\right\} = \frac{t}{4}\sinh(t)$$

▶

EXAMPLE 3.12

Solve the differential equation

$$t \frac{d^2 x}{dt^2} + at \frac{dx}{dt} + ax = 0 \tag{3.1}$$

given that $x = 0$, $dx/dt = A$ when $t = 0$.

Note that all differential equations considered previously have involved constant coefficients. In this case the method remains as follows: transform the equation, solve the resulting equation and then invert. Hence, transforming equation (3.1),

$$-\frac{d}{ds}(s^2 \bar{x} - sx_0 - x_0') - a \frac{d}{ds}(s\bar{x} - x_0) + a\bar{x} = 0$$

Since $x_0 = 0$, $x_0' = A$,

$$-\frac{d}{ds}(s^2 \bar{x} - A) - a \frac{d}{ds}(s\bar{x}) + a\bar{x} = 0$$

The terms $s^2 \bar{x}$ and $s\bar{x}$ are products of functions of s and must be differentiated as such.

$$-2s\bar{x} - s^2 \frac{d\bar{x}}{ds} - a\left(\bar{x} + s \frac{d\bar{x}}{ds}\right) + a\bar{x} = 0$$

$$\frac{d\bar{x}}{ds}(-s^2 - as) - 2s\bar{x} = 0$$

$$\frac{d\bar{x}}{ds}(s + a) + 2\bar{x} = 0 \tag{3.2}$$

The resulting equation (3.2) is therefore another differential equation but in $\bar{x}$ and s which may or may not be easier to solve than the original equation (3.1).

In this case the variables $\bar{x}$ and s can be separated.

$$\frac{d\bar{x}}{\bar{x}} = -\frac{2\,ds}{s+a}$$

$$\ln(\bar{x}) = -2\ln(s+a) + K$$

$$\bar{x} = \frac{C}{(s+a)^2}$$

where $C = \ln(K)$. Inverting, the solution is

$$x = Cte^{-at}$$

Periodic functions (9)

Suppose that $f(t)$ is a periodic function of period T so that $f(t + T) = f(t)$ for all values of t. Then

$$\mathscr{L}\{f(t)\} = \int_0^\infty e^{-st} f(t) \, dt$$

$$= \int_0^T e^{-st} f(t) \, dt + \int_T^{2T} e^{-st} f(t) \, dt$$

$$+ \int_{2T}^{3T} e^{-st} f(t) \, dt + \cdots$$

$$= \int_0^T e^{-st} f(t) \, dt + e^{-sT} \int_0^T e^{-st} f(t) \, dt$$

$$+ e^{-2sT} \int_0^T \cdots$$

$$= \{1 + e^{-sT} + e^{-2sT} + \cdots\} \int_0^T e^{-st} f(t) \, dt$$

Since $1 + e^{-sT} + e^{-2sT} + \cdots$ is an infinite geometric series, its sum to infinity is $1/(1 - e^{-sT})$ provided that $e^{-sT} < 1$. The transform of $f(t)$ is then given by

$$\mathscr{L}\{f(t)\} = \frac{1}{1 - e^{-sT}} \int_0^t e^{-sT} f(t) \, dt$$

EXAMPLE 3.13

Determine the Laplace transform of the square wave defined by

$$f(t) = \begin{cases} E & 0 \leqslant t < T/2 \\ -E & T/2 \leqslant t < T \end{cases}$$

$$f(t + T) = f(t)$$

$$F(s) = \frac{1}{1 - e^{-sT}} \int_0^T e^{-st} f(t) \, dt$$

$$= \frac{1}{1 - e^{-sT}} \left\{ \int_0^{T/2} E e^{-st} \, dt - \int_{T/2}^T E e^{-st} \, dt \right\}$$

$$= \frac{E}{1 - e^{-sT}} \left\{ \frac{1}{s} \left[1 - e^{-sT/2} - e^{-sT/2} + e^{-sT} \right] \right\}$$

$$= \frac{E}{s(1 - e^{-sT})} (1 - e^{-sT/2})^2$$

$$= \frac{E(1 - e^{-sT/2})}{s(1 + e^{-sT/2})}$$

since $1 - e^{-sT} = (1 - e^{-sT/2})(1 + e^{-sT/2})$. From the definition of the hyperbolic tangent this result may also be written as

$$F(s) = \frac{E}{s} \tanh \left(\frac{sT}{4} \right)$$

EXAMPLE 3.14

Solve the equation

$$\frac{d^2 x}{dt^2} + \omega^2 x = f(t)$$

where $f(t)$ is the square wave defined in Example 3.13 and given that $x = dx/dt = 0$ when $t = 0$.

Transforming, we have

$$(s^2 \bar{x} - s x_0 - x_0') + \omega^2 \bar{x} = \mathcal{L}\{f(t)\}$$

$$\bar{x}(s^2 + \omega^2) = \frac{1}{s} \frac{1 - e^{-sT/2}}{1 + e^{-sT/2}}$$

Therefore

$$\bar{x} = \frac{1}{s(s^2 + \omega^2)} \frac{1 - e^{-sT/2}}{1 + e^{-sT/2}}$$

The only standard form which might enable us to deal with $F(s)$ multiplied by an exponential is standard form 8 of Appendix 2 which is yet to be discussed. We shall therefore leave the formal inverse until later; however, we need the exponential function to be in the numerator. Therefore

$$\bar{x} = \frac{1}{s(s^2 + \omega^2)} (1 - e^{-sT/2})(1 + e^{-sT/2})^{-1}$$

$$= \frac{1}{s(s^2 + \omega^2)} (1 - e^{-sT/2})(1 - e^{-sT/2} + e^{-sT} - e^{-3sT/2} + \cdots)$$

$$= \frac{1}{s(s^2 + \omega^2)} (1 - 2e^{-sT/2} + 2e^{-sT} - 2e^{-3sT/2} + \cdots)$$

Initial value theorem

If the indicated limits exist, then

$$\lim_{t \to 0} f(t) = \lim_{s \to \infty} sF(s)$$

PROOF $\mathcal{L}\{f'(t)\} = \int_0^\infty e^{-st} f'(t) \, dt$. Integration by parts gives

$$\mathcal{L}\{f'(t)\} = [e^{-st} f(t)]_0^\infty - \int_0^\infty - se^{-st} f(t) \, dt$$

$$= -f(0) + sF(s)$$

If $f'(t)$ is sectionally continuous and of exponential order then

$$\lim_{s \to \infty} \int_0^\infty e^{-st} f'(t) \, dt = 0$$

Therefore

$$\lim_{s \to \infty} [sF(s) - f(0)] = 0$$

$$\lim_{s \to \infty} sF(s) = f(0) = \lim_{t \to 0} f(t)$$

If $f(t)$ is not continuous at $t = 0$ the required result still holds.

Final value theorem

If the indicated limits exist, then

$$\lim_{t \to \infty} f(t) = \lim_{s \to 0} sF(s)$$

PROOF From above

$$\mathcal{L}\{f'(t)\} = sF(s) - f(0)$$

$$\lim_{s \to 0} \mathcal{L}\{f'(t)\} = \lim_{s \to 0} \int_0^\infty e^{-st} f'(t) \, dt$$

$$= \int_0^\infty f'(t) \, dt$$

$$= \lim_{t \to \infty} \int_0^t f'(t) \, dt$$

$$= \lim_{t \to \infty} f(t) - f(0)$$

The limit of the right-hand side is

$$\lim_{s \to 0} sF(s) - f(0)$$

Hence

$$\lim_{t \to \infty} f(t) = \lim_{s \to 0} sF(s)$$

In many applications of these theorems $\mathscr{L}\{f(t)\}$ will be known but $f(t)$ will not. It can also be shown that the final value theorem cannot be applied if there is any value of s with non-negative real part for which $sF(s)$ is unbounded. For example note that

$$\lim_{s \to 0} \frac{s}{s^2 + 1} = 0$$

The theorem cannot be applied to $F(s) = 1/(s^2 + 1)$ since this is unbounded for the value $s = \pm j$. In this case $f(t) = \mathscr{L}^{-1}\{F(s)\} = \sin(t)$ and $\lim_{t \to \infty} \sin(t)$ does not exist.

In the modeling and control of systems it is useful to obtain the values of the steady state gain and the steady state error. Examples of these are given after discussion of transfer functions in Chapter 6 but for the moment two examples follow to indicate other uses of these theorems.

EXAMPLE 3.15

Determine $\bar{y}$ if

$$y = \int_t^\infty \frac{e^{-u}}{u} \, du$$

Since

$$y = \int_t^\infty \frac{e^{-u}}{u} \, du$$

then

$$t \frac{dy}{dt} = -e^{-t}$$

Transforming we have

$$-\frac{d}{ds} [s\bar{y} - y(0)] = -\frac{1}{s + 1}$$

$$\frac{d}{ds} (s\bar{y}) = \frac{1}{s + 1}$$

$$s\bar{y} = \ln(s + 1) + A$$

Applying the final value theorem, both limits must be zero. Therefore $A = 0$. Hence

$$\bar{y} = \frac{1}{s} \ln(s + 1)$$

EXAMPLE 3.16

If $s\bar{y} = - \tan^{-1}(s) + C$ and $y = 0$ when $t = 0$ determine y.

$$s\bar{y} = - \tan^{-1}(s) + C$$

By the initial value theorem

$$\lim_{s \to \infty} s\bar{y} = \lim_{t \to 0} y = 0$$

Therefore $C = \pi/2$. Hence

$$s\bar{y} = - \tan^{-1}(s) + \frac{\pi}{2} = \tan^{-1}\left(\frac{1}{s}\right)$$

Since

$$\mathscr{L}^{-1}\left\{\frac{1}{s} F(s)\right\} = \int_0^t \mathscr{L}^{-1}\{F(s)\} \; ds$$

$$y = \mathscr{L}^{-1}\left\{\frac{1}{s} \tan^{-1}\left(\frac{1}{s}\right)\right\} = \int_0^t \mathscr{L}^{-1}\left\{\tan^{-1}\left(\frac{1}{s}\right)\right\} \; dt$$

Since

$$\mathscr{L}^{-1}\{F(s)\} = -\frac{1}{t} \mathscr{L}^{-1}\left\{\frac{d}{ds} [F(s)]\right\}$$

$$y = \int_0^t -\frac{1}{t} \mathscr{L}^{-1}\left\{\frac{d}{ds}\left[\tan^{-1}\left(\frac{1}{s}\right)\right]\right\} \; dt$$

$$= \int_0^t -\frac{1}{t} \mathscr{L}^{-1}\left\{-\frac{1}{1 + s^2}\right\} \; dt$$

$$= \int_0^t \frac{1}{t} \sin(t) \; dt$$

EXERCISE 3.1

Determine the Laplace transforms of each of the following functions where ω, a, b, k and T are constant:

1 $t \sin(\omega t)$ **2** $t^2 \cos(\omega t)$

3 $t\{6 \sin(t) - 8 \cos(t)\}$

4 $t \cosh(2t)$

5 $t \sinh(4t)$

6 $\int_0^t te^{-2t} \sin(3t)\, dt$

7 $\int_0^t \frac{1}{t} e^{-2t} \sin(3t)\, dt$

8 $\frac{1}{t} \{e^{-bt} - e^{-at}\}$

9 $\frac{1}{t} \sinh(t)$

10 $f(t) = \begin{cases} \sin(t) & 0 \leqslant t < \pi \\ 0 & \pi \leqslant t < 2\pi \end{cases}$

Period 2π

11 $f(t) = \begin{cases} t & 0 \leqslant t < 1 \\ 0 & 1 \leqslant t < 2 \end{cases}$

Period 2

12 $y = kt \qquad 0 < t \leqslant T$

Period T

Determine the inverse of the following.

13 $\int_s^\infty \frac{1}{s(s+1)}\, ds$

14 $\ln\left(1 + \frac{1}{s}\right)$

15 $\frac{1}{s} \ln\left(1 + \frac{1}{s}\right)$

16 $\ln\left(\frac{s+a}{s+b}\right)$

17 $\frac{1}{s} \ln\left(\frac{s+a}{s+b}\right)$

18 $\frac{1}{s^2(s+1)}$

19 $\frac{1}{s(s+1)^2}$

4

The Heaviside step function

Introduction

The Heaviside step function (unit step function) denoted by $u(t)$ or $H(t)$ is defined as (Figure 4.1)

$$u(t) = \begin{cases} 0 & t < 0 \\ 1 & t \geqslant 0 \end{cases}$$

If T is any fixed constant then (Figure 4.2)

$$u(t - T) = \begin{cases} 0 & t < T \\ 1 & t \geqslant T \end{cases}$$

i.e. the Heaviside function shifted by T units.

The function is defined such that the equality appears once and once

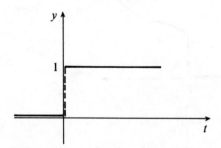

Figure 4.1   The unit step function $y = u(t)$

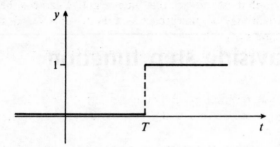

Figure 4.2 The delayed step function $y = u(t - T)$

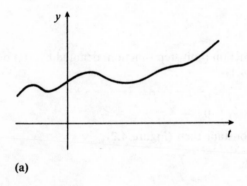

(a)

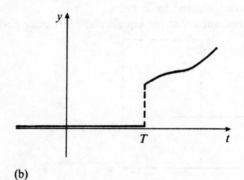

(b)

Figure 4.3 (a) The function $y = f(t)$; (b) the function $y = f(t)u(t - T)$

only. From the graph it can be seen that an alternative name is the 'unit step function' which may be interpreted as a model of a switch that is switched on at time $t = T$.

The combination of the step function and any other function $f(t)$, i.e.

$$f(t)u(t - T)$$

is clearly a function $f(t)$ multiplied by 0 if $t < T$, 1 if $t \geqslant T$, i.e. a function which is equal to 0 if $t < T$ and $f(t)$ if $t \geqslant T$. This combination is shown in Figure 4.3.

For $T \geqslant 0$, the Laplace transform of the unit step function is

$$\mathscr{L}\{u(t - T)\} = \int_0^\infty e^{-st} u(t - T) \, dt$$

$$= \int_0^T 0 \, dt + \int_T^\infty e^{-st} \, dt$$

$$= \frac{e^{-sT}}{s}$$

This result is listed as 11 in Appendix 2.

EXAMPLE 4.1

Sketch the function

$$y = 3u(t - 1) - 2u(t - 2)$$

and determine its Laplace transform.

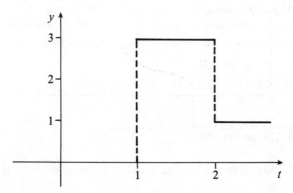

Figure 4.4   The function $y = 3u(t - 1) - 2u(t - 2)$

The function represents a step of magnitude 3 at $t = 1$ followed by a step of magnitude -2 at $t = 2$. The sketch is therefore that shown in Figure 4.4.

$$\mathscr{L}\{3u(t-1) - 2u(t-2)\} = 3\mathscr{L}\{u(t-1)\} - 2\mathscr{L}\{u(t-2)\}$$
$$= \frac{3e^{-s}}{s} - \frac{2e^{-2s}}{s}$$

using the above result.

EXAMPLE 4.2

Determine an expression for the function

$$f(t) = \begin{cases} 5 & a \leqslant t < b \\ 0 & \text{elsewhere} \end{cases}$$

This function is considered in two parts: the operation at $t = a$ (switch on), and the operation at $t = b$ (switch off). At $t = a$ there is a step of magnitude $+5$, i.e. $5u(t-a)$. At $t = b$ there is a step of magnitude -5, i.e. $-5u(t-b)$. Therefore

$$f(t) = 5u(t-a) - 5u(t-b)$$

EXAMPLE 4.3

Determine an expression for

$$f(t) = \begin{cases} t-1 & 1 \leqslant t < 3 \\ 8-2t & 3 \leqslant t < 4 \end{cases}$$

There are four operations to consider: at $t = 1, t = 3$ (two) and $t = 4$.

at $t = 1$, $t - 1$ is introduced	$(t-1)u(t-1)$
at $t = 2$, $t - 1$ is removed	$-(t-1)u(t-3)$
at $t = 2$, $8 - 2t$ is introduced	$(8-2t)u(t-3)$
at $t = 4$, $8 - 2t$ is removed	$-(8-2t)u(t-4)$

$$f(t) = (t-1)u(t-1) - (t-1)u(t-3) + (8-2t)u(t-3)$$
$$- (8-2t)u(t-4)$$
$$= (t-1)u(t-1) - 3(t-3)u(t-3) - (8-2t)u(t-4)$$

Second shift theorem (8)

The second shifting property of the Laplace transform states that if $\mathscr{L}\{f(t)\}$ exists then

$$\mathscr{L}\{f(t-T)u(t-T)\} = e^{-sT}F(s)$$

or, in inverse form,

$$\mathcal{L}^{-1}\{e^{-sT}F(s)\} = f(t - T)u(t - T)$$

The function $f(t)$ rarely appears in the form $f(t - T)$. It is necessary either to rewrite the standard form as

$$\mathcal{L}\{f(t)u(t - T)\} = e^{-sT}\mathcal{L}\{f(t + T)\}$$

or to rewrite $f(t)$ in the form $f(t - T)$. The latter is less suitable for dealing with an inverse so that a rewrite of $f(t)$, when necessary, is the preferred method. It is again important to be aware of the relationship $\mathcal{L}\{f(t)\} = F(s)$ and the meaning of $f(t - T)$. For example if $f(t) = \sin(2t)$ then $f(t - T) = \sin[2(t - T)]$.

EXAMPLE 4.4

Determine the Laplace transform of $(t - 2)u(t - 2)$.

$$\mathcal{L}\{(t - 2)u(t - 2)\} = e^{-2s}\mathcal{L}\{t\} = e^{-2s}\frac{1}{s^2}$$

since $T = 2$ and $f(t - 2) = t - 2$, so that $f(t) = t$ and $F(s) = 1/s^2$.

EXAMPLE 4.5

Find the Laplace transform of

$$f(t) = \begin{cases} 0 & t < T \\ -e^{-t} & t \geqslant T \end{cases}$$

$$f(t) = -e^{-t}u(t - T) = -e^{-(t - T + T)}u(t - T)$$
$$= -e^{-T}e^{-(t - T)}u(t - T)$$

Note that t is written as $t - T + T$ and that the laws of indices are used to write $e^{-(t - T + T)}$ as $e^{-(t - T)}e^{-T}$.

$$\mathcal{L}\{f(t)\} = \mathcal{L}\{-e^{-T}e^{-(t - T)}u(t - T)\}$$

$$= -e^{-T}e^{-sT}\frac{1}{s + 1}$$

$$= \frac{-e^{-(s + 1)T}}{s + 1}$$

using 8 and 17 of Appendix 2.

EXERCISE 4.1

Sketch the graph of each of the following functions and find their Laplace transforms:

1 $f(t) = 5 - u(t - 1)$ **2** $f(t) = 4 - 2u(t - 2)$

3 $f(t) = u(t - 1) + 3u(t - 4)$ **4** $f(t) = 2u(t) - u(t - 3)$

5 $f(t) = u(t - 1) - 2u(t - 2) + 4u(t - 3)$

Write the following as a single equation using the unit step function; sketch the graphs and find the Laplace transforms:

6 $y = 0$ when $t < 3$, $y = 5$ when $t \geqslant 3$

7 $y = 3$ when $t < 1$, $y = -1$ when $t \geqslant 1$

8 $y = 0$ when $t < 1$, $y = 2$ when $1 \leqslant t < 3$, $y = -2$ when $t \geqslant 3$

Express the following as a single equation using the step function and find the Laplace transform:

9

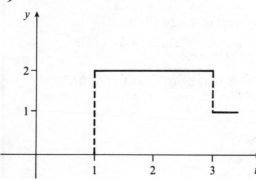

10

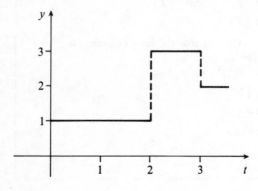

11

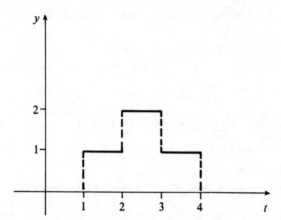

12 Show that
$$f(t) = \begin{cases} e^{-3t} & 0 < t \leqslant 2 \\ 0 & t > 2 \end{cases}$$
can be written as $e^{-3t}[1 - u(t - 2)]$ and hence find $\mathscr{L}\{f(t)\}$.

13 Express the following in terms of Heaviside's unit step function:

(a) $f(t) = \begin{cases} t & 0 < t \leqslant 1 \\ -t + 2 & t > 1 \end{cases}$

(b) $f(t) = \begin{cases} \cos(t) & 0 \leqslant t < \pi \\ \cos(2t) & \pi \leqslant t < 2\pi \\ \cos(3t) & 2\pi \leqslant t \end{cases}$

Find the Laplace transforms of the following:

14 $2tu(t - 3)$

15 $e^{3t}u(t - 1)$

16 $f(t) = \begin{cases} \cos(t - \pi/3) & t \geqslant \pi/3 \\ 0 & \text{elsewhere} \end{cases}$

17 $f(t) = \begin{cases} (t - 1)^3 & t \geqslant 1 \\ 0 & \text{elsewhere} \end{cases}$

18 $f(t) = \begin{cases} e^{3t} & t \geqslant 1 \\ 0 & \text{elsewhere} \end{cases}$

19 A square wave is given in the form

$$y = \begin{cases} E & 0 \leqslant t < T \\ -E & T \leqslant t < 2T \\ E & 2T \leqslant t < 3T \text{ etc.} \end{cases}$$

Show that

$$\bar{y} = \frac{E}{s} \tanh(sT/2).$$

20 Determine $\bar{y}$ if $y = 1, 2, 3, \ldots$ for the intervals $(0, T)$, $(T, 2T)$, $(2T, 3T), \ldots$ (the staircase function).

EXAMPLE 4.6

Determine

$$\mathscr{L}^{-1}\left\{ e^{-as}\left(\frac{1}{s+1} - \frac{1}{s+2} \right) \right\}$$

From the table of standard forms (Appendix 2, number 17)

$$\mathscr{L}^{-1}\left\{ \frac{1}{s+1} \right\} = e^{-t} \qquad \mathscr{L}^{-1}\left\{ \frac{1}{s+2} \right\} = e^{-2t}$$

(ignoring e^{-as}) since it is necessary to determine $f(t)$ the inverse of $F(s)$.

$$\mathscr{L}^{-1}\left\{ e^{-as}\left(\frac{1}{s+1} - \frac{1}{s+2} \right) \right\} = e^{-(t-a)}u(t-a) - e^{-2(t-a)}u(t-a)$$

EXAMPLE 4.7

Solve the differential equation

$$\frac{dy}{dt} + 3y = f(t)$$

where

$$f(t) = \begin{cases} 0 & t < 1 \\ 2 & 1 \leqslant t < 2 \\ 0 & t \geqslant 2 \end{cases}$$

and $y = 1$ when $t = 0$.

i.e. $f(t) = 2u(t-1) - 2u(t-2)$

Transforming the equation gives

$$s\bar{y} - y_0 + 3\bar{y} = \frac{2e^{-s}}{s} - \frac{2e^{-2s}}{s}$$

$$\bar{y} = \frac{1}{s+3} + \frac{2e^{-s}}{s(s+3)} - \frac{2e^{-2s}}{s(s+3)}$$

Since

$$\mathscr{L}^{-1}\left\{\frac{1}{s+3}\right\} = e^{-3t} \qquad \mathscr{L}^{-1}\left\{\frac{1}{s(s+3)}\right\} = \tfrac{1}{3}(1 - e^{-3t})$$

the inverse of $\bar{y}$ is

$$y = e^{-3t} + \tfrac{2}{3}[1 - e^{-3(t-1)}]u(t-1) - \tfrac{2}{3}[1 - e^{-3(t-2)}]u(t-2)$$

If it is necessary to use partial fractions in order to determine the inverse of $F(s)$ then the factor e^{-as} must not be included in that calculation. For example in calculating $\mathscr{L}^{-1}\{e^{-s}/(s+1)(s+2)\}$

$$F(s) = \frac{1}{(s+1)(s+2)} \qquad f(t) = \mathscr{L}^{-1}\{F(s)\}$$

$$\mathscr{L}^{-1}\left\{\frac{1}{(s+1)(s+2)}\right\} = \mathscr{L}^{-1}\left\{\frac{1}{s+1} - \frac{1}{s+2}\right\} = e^{-t} - e^{-2t}$$

Hence

$$\mathscr{L}^{-1}\left\{\frac{e^{-s}}{(s+1)(s+2)}\right\} = [e^{-(t-1)} - e^{-2(t-1)}]u(t-1)$$

EXAMPLE 4.8

The solution of Example 3.14 of a square wave forcing function now becomes

$$\bar{x} = \frac{1}{s(s^2 + \omega^2)} \frac{1 - e^{-sT/2}}{1 + e^{-sT/2}}$$

$$= \frac{1}{s(s^2 + \omega^2)} (1 - e^{-sT/2})(1 + e^{-sT/2})^{-1}$$

$$= \frac{1}{s(s^2 + \omega^2)} (1 - 2e^{-sT/2} + 2e^{-sT} - 2e^{-3sT/2} + \cdots)$$

▶

using a binomial expansion. Since

$$\mathscr{L}^{-1}\left\{\frac{1}{s(s^2 + \omega^2)}\right\} = \tfrac{1}{2}[1 - \cos(\omega t)]$$

$$x = \tfrac{1}{2}\{[1 - \cos(\omega t)] - 2[1 - \cos \omega(t - T/2)]u(t - T/2)$$
$$+ 2[1 - \cos \omega(t - T)]u(t - T) - \cdots\}$$

EXAMPLE 4.9

A light horizontal beam of length l is clamped at each end and carries a load w N m^{-1} over the length $x = a$ to $x = b$ measured from the left-hand end of the beam. Determine the deflection of the beam.

The mathematical model of this is the differential equation

$$EI\frac{\mathrm{d}^4 y}{\mathrm{d}x^4} = wu(x - a) - wu(x - b)$$

subject to the conditions $y = 0$, $\mathrm{d}y/\mathrm{d}x = 0$ at $x = 0$ and $y = 0$, $\mathrm{d}y/\mathrm{d}x = 0$ at $x = l$ (i.e. zero deflection and zero gradient at each end).

Taking the Laplace transform of the equation gives

$$EI\{s^4\bar{y} - s^3 y_0 - s^2 y_0' - s y_0'' - y_0'''\} = w\left(\frac{e^{-as}}{s} - \frac{e^{-bs}}{s}\right)$$

At $x = 0$, $y = 0$, $\mathrm{d}y/\mathrm{d}x = 0$ but $\mathrm{d}^2y/\mathrm{d}x^2$ and $\mathrm{d}^3y/\mathrm{d}x^3$ are unknown constants.

Let $\mathrm{d}^2y/\mathrm{d}x^2 = A$ and $\mathrm{d}^3y/\mathrm{d}x^3 = B$ at $x = 0$. The equation becomes

$$EI\{s^4\bar{y} - As - B\} = w\left(\frac{e^{-as}}{s} - \frac{e^{-bs}}{s}\right)$$

and solving for $\bar{y}$ we obtain

$$\bar{y} = \frac{A}{s^3} + \frac{B}{s^4} + \frac{w}{EI}\left(\frac{e^{-as}}{s^5} - \frac{e^{-bs}}{s^5}\right)$$

which on using 16 and 8 of Appendix 2 for inversion gives

$$y = \frac{A}{2}x^2 + \frac{B}{6}x^3 + \frac{w}{24EI}[(x - a)^4 u(x - a) - (x - b)^4 u(x - b)] \tag{4.1}$$

The remaining conditions $y = 0$, $\mathrm{d}y/\mathrm{d}x = 0$ at $x = l$ are now used to determine A and B. On substitution of $y = 0$, $x = l$ into equation (4.1) we obtain

$$0 = \frac{A}{2}l^2 + \frac{B}{6}l^3 + \frac{w}{24EI}[(l - a)^4 - (l - b)^4] \tag{4.2}$$

and by differentiation of equation (4.1) with respect to x and substitution of $dy/dx = 0$ at $x = l$ we have

$$0 = Al + \frac{B}{2} l^2 + \frac{w}{6EI} [(l-a)^3 - (l-b)^3] \tag{4.3}$$

Note that $u(x - a) = u(x - b) = 1$ at $x = l$. Hence the constants A and B are obtained by solving the simultaneous equations (4.2) and (4.3).

Some properties of $u(t)$

1. The solution of many problems begins at $t = 0$ so that any analysis for $t < 0$ is of no interest

    $$u(t) = 1 \qquad t \geqslant 0$$

2. The step function can be treated algebraically in the same way as any other function. It is a multiplying factor which is zero for $t < 0$ and 1 for $t \geqslant 0$.

3. $u(-t) = 1 - u(t)$ (Figure 4.5).

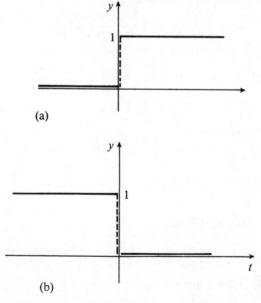

(a)

(b)

Figure 4.5 (a) The step function $y = u(t)$; (b) the function $y = u(-t)$

4. $u(T-t) = u[-(t-T)] = 1 - u(t-T)$.
5. Scaling: $u(at) = u(t)$

   i.e. the step at $t = 0$ is independent of a;

$$u(at - T) = u(t - T/a)$$

   i.e. step occurs when $at = T$, $t = T/a$.

It may be possible to sum an infinite series after transforming each term. For example in transforming a square wave function or a staircase function the resulting series is a geometric series as in Chapter 3. The sum to infinity of a geometric series is

$$S = a + ar + ar^2 + ar^3 + \cdots = a/(1 - r)$$

provided that $|r| < 1$.

EXERCISE 4.2

Find the inverse of the following transforms:

1 $\dfrac{se^{-s}}{s^2 + 9}$

2 $\dfrac{e^{-s}s^2}{(s-2)(s^2+4)}$

3 $\dfrac{e^{-2s}}{(s+1)^4} - \dfrac{e^{-s}s}{(s^2+4)}$

4 $\dfrac{e^{-3s}}{s^2 + 9}$

5 $\dfrac{e^{-s}}{(s+2)^3}$

6 $\dfrac{8e^{-2s}}{s^2 + 4}$

7 $\dfrac{e^{-\pi s/6}}{s^2 - 1}$

8 $\dfrac{e^{-5s}}{(s-3)^4}$

9 $\dfrac{e^{-2s} - e^{-3s}}{s(s^2 + 1)}$

Solve the following differential equations:

10 $\dfrac{d^2i}{dt^2} - 4\dfrac{di}{dt} - 5i = 30u(t-1)$

   given that $i = 0$, $di/dt = 6$ at $t = 0$.

11 $\dfrac{d^2x}{dt^2} + 3\dfrac{dx}{dt} + 2x = u(t-2)$

   given that $x = 0$, $dx/dt = 0$ at $t = 0$.

12 $\dfrac{d^2x}{dt^2} + x = f(t)$

where $f(t) = 2$, $0 < t < 3$, and $x = 1$, $dx/dt = 0$ at $t = 0$.

13 $\dfrac{d^2y}{dt^2} + 4\dfrac{dy}{dt} + 4y = 8e^{-(t-1)}u(t-1)$

given that $y = -1$, $dy/dt = 1$ when $t = 0$.

14 The voltage E applied to an electrical circuit is given by
$$E = \begin{cases} 25t & 0 \leqslant t < 2 \\ 50 & t \geqslant 2 \end{cases}$$
Determine the current in the circuit if both current and charge are initially zero, i.e. solve the equations
$$\frac{di}{dt} + 16q = E \qquad i = \frac{dq}{dt}$$
if $i = 0$ and $q = 0$ when $t = 0$.

15 The external force $f(t)$ acting on a mechanical system is given by
$$f(t) = \begin{cases} \cos(2t) & 0 \leqslant t < \pi \\ 0 & t \geqslant \pi \end{cases}$$
Solve the differential equation
$$\frac{d^2x}{dt^2} + 4x = f(t)$$
given that $x = 0$, $dx/dt = 3$ when $t = 0$.

16 The currents in an electrical circuit are given by the equations
$$L\frac{di_1}{dt} + L\frac{di_2}{dt} + R_1 i_1 = E$$

$$-R_1\frac{di_1}{dt} + R_2\frac{di_2}{dt} + \frac{1}{C}i_2 = 0$$
If $R_1 = 5$ ohms, $R_2 = 10$ ohms, $L = 1$ henry and $C = 0.1$ farads determine i_2 if $E = 100$ volts for $0 < t \leqslant 1$ and initially both i_1 and i_2 are zero.

5

The impulse function

Introduction

It is convenient to have a notation for intense pulses so brief that measuring equipment of a given resolving power is unable to distinguish between them and even briefer pulses. This concept is covered in mechanics by the term *impulse*. The idea has been in use for more than a century in mathematical circles and was extensively employed by Heaviside. The notation $\delta(t)$, introduced by Dirac into quantum mechanics, is now in general use.

We are familiar with the ideas of point masses, point charges, point sources, concentrated forces, surface charges, etc., which are accepted entities. These things do not actually exist. Their value stems from the fact that the impulse response – the effect associated with the impulse – may be indistinguishable, given measuring equipment of specified resolving power, from the response due to a physically realizable pulse. It is thus convenient to have a name for pulses which are so brief and intense that making them any briefer or more intense does not matter.

For example, if we consider a mass (or charge) distribution along the x axis density $\rho(x)$ then

$$\text{total mass (charge) between } a \text{ and } b = \int_a^b \rho(x) \, dx$$

The idea of a point mass (unit point charge) concentrated at $x = 0$ is then a limiting process, a more and more intense distribution over a narrower and narrower band. The delta function $\delta(x)$ is defined as the limiting density for the total point mass (charge) to be equal to unity, i.e.

$$\int_{-\infty}^{\infty} \delta(x) \, \mathrm{d}x = 1$$

However, $\delta(x) = 0$ for all x except $x = 0$ and $\delta(0)$ must be infinite. For a particle of unit mass at x_0, the corresponding density is $\delta(x - x_0)$. Although the impulse function and associated functions, referred to as generalized functions, are idealized functions, the operations involving them can be made mathematically rigorous. These functions can be combined algebraically, integrated and differentiated. Their value in the analysis of engineering and physical systems is considerable.

A more general property is that

$$\int_{-\infty}^{\infty} f(x) \, \delta(x - x_0) \, \mathrm{d}x = f(x_0)$$

which will be discussed later.

The graphical representation of the delta function is shown in Figures 5.1 and 5.2. The function may also be defined as

$$\delta(t) = \lim_{\varepsilon \to 0} \frac{1}{\varepsilon} \, [u(t) - u(t - \varepsilon)] = \frac{\mathrm{d}}{\mathrm{d}t} \, [u(t)]$$

i.e. the derivative of the step function. The Laplace transforms are

$$\mathscr{L}\{\delta(t)\} = \mathscr{L}\left\{\frac{\mathrm{d}}{\mathrm{d}t} \, u(t)\right\} = \frac{1}{s} \, \mathscr{L}\{u(t)\} = \frac{1}{s} \, s = 1$$

$$\mathscr{L}\{\delta(t - a)\} = \mathrm{e}^{-as} \mathscr{L}\{\delta(t)\} = \mathrm{e}^{-as}$$

by the shift theorem.

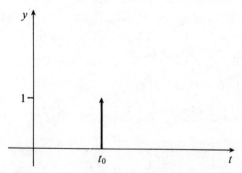

Figure 5.1   The impulse function $y = \delta(t - t_0)$

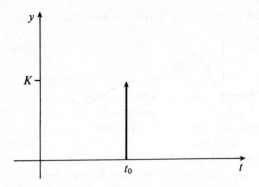

Figure 5.2 The impulse of strength K, $y = K\delta(t - t_0)$

Other generalized functions are as follows:

unit doublet (dipole) $\delta'(t)$ $\mathscr{L}\{\delta'(t)\} = s$
unit triplet $\delta''(t)$ $\mathscr{L}\{\delta''(t)\} = s^2$

Some important properties

1. $\displaystyle\int_{-\infty}^{\infty} \delta(t)\, dt = 1$ $\displaystyle\int_{-\infty}^{\infty} \delta(t - a)\, dt = 1$

2. Since the area (integral) is equal to unity we define the strength of the impulse as being equal to unity. Hence an impulse of magnitude K occurring at $t = a$ is written as $K\,\delta(t - a)$.

3. (a) $\delta(t)f(t) = f(t)\,\delta(t) = f(0)\,\delta(t)$
   (b) $\delta(t - a)f(t) = f(a)\,\delta(t - a)$

These rules are suggested by the interpretation of $\delta(t)$, $\delta(t - a)$ as pulses so that the product of the two functions is zero for all values of t except that at which the impulse occurs.

4. $\displaystyle\int_{-\infty}^{\infty} f(t)\,\delta(t - a)\, dt = f(a)$

Note that $f(a)$ is a constant – the numerical value of $f(t)$ at $t = a$ (Figure 5.3).

The integral gives the area under the curve of the combined functions. This is the sifting property of the impulse function since the operation on $f(t)$ sifts out a single value of $f(t)$.

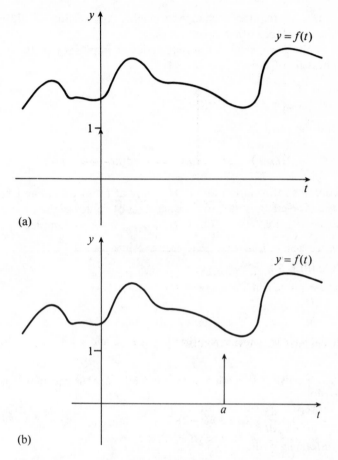

(a)

(b)

Figure 5.3 (a) The combination $y = f(t)$ and $y = \delta(t)$; (b) the combination $y = f(t)$ and $y = \delta(t - a)$

5. $$\int_a^b \delta(t)\ \mathrm{d}t = 1$$

if the range of integration includes $t = t_0$; otherwise the integral is zero.

6. Scaling:

$$\delta(at) = \frac{1}{|a|}\ \delta(t) \qquad \delta(at - t_0) = \frac{1}{|a|}\ \delta(t - t_0/a)$$

Areas under the graphs must be equal; therefore a change in the time scale requires a change in height. Any scaling should be considered before other operations.

7. Sampling property: An infinite series of impulses equally spaced at intervals of t_0 is

$$\sum_{n=-\infty}^{\infty} \delta(t - nt_0)$$

and the product

$$f(t) \sum_{n=-\infty}^{\infty} \delta(t - nt_0) = \sum_{n=-\infty}^{\infty} f(nt_0) \delta(t - nt_0)$$

contains information about $f(t)$ only at times $t = nt_0$ and *nowhere else*. The effect is then to produce a set of values of the function $f(t)$ at $t = nt_0$, i.e. $f(t_0)$, $f(2t_0)$, $f(3t_0)$, ..., a set of sampled values.

EXAMPLE 5.1

Solve the differential equation

$$\frac{d^2x}{dt^2} + 3 \frac{dx}{dt} + 2x = f(t)$$

where $f(t)$ is an impulse of strength 5 at $t = 2$ and $x = 4$, $dx/dt = 0$ at $t = 0$.

An impulse of strength 5 at $t = 2$ is $5\,\delta(t - 2)$. The equation to be solved is

$$\frac{d^2x}{dt^2} + 3 \frac{dx}{dt} + 2x = 5\,\delta(t - 2)$$

Transforming we have

$$(s^2\bar{x} - sx_0 - x_0') + 3(s\bar{x} - x_0) + 2\bar{x} = 5e^{-2s}$$

Since $x_0 = 4$, $x_0' = 0$,

$$\bar{x}(s^2 + 3s + 2) = 5e^{-2s} + 4s + 12$$

$$\bar{x} = \frac{1}{(s+1)(s+2)} (5e^{-2s} + 4s + 12)$$

From Appendix 2

$$\mathscr{L}^{-1}\left\{\frac{1}{(s+1)(s+2)}\right\} = e^{-t} - e^{-2t}$$

and

$$\mathscr{L}^{-1}\left\{\frac{s}{(s+1)(s+2)}\right\} = 2e^{-2t} - e^{-t}$$

Therefore

$$x = 5[e^{-(t-2)} - e^{-2(t-2)}]u(t-2) + 4(2e^{-2t} - e^{-t})$$
$$+ 12(e^{-t} - e^{-2t})$$
$$= 5[e^{-(t-2)} - e^{-2(t-2)}]u(t-2) + 8e^{-t} - 4e^{-2t}$$

EXAMPLE 5.2

Solve for i the equations

$$L\frac{di}{dt} + \frac{q}{C} = E \qquad i = \frac{dq}{dt}$$

where E is an impulse of magnitude E_0 at $t = 0$ and initially both current i and charge q are zero.

E is therefore given by $E = E_0 \delta(t)$, and so the transformed equations are

$$L(s\bar{i} - i_0) + \bar{q}/C = \mathscr{L}\{E_0 \delta(t)\} = E_0$$
$$\bar{i} = s\bar{q} - q_0$$

Eliminating $\bar{q}$ we have

$$\bar{i}(Ls + 1/Cs) = E_0$$
$$\bar{i} = \frac{E_0 s}{Ls^2 + 1/C} = \frac{E_0}{L}\frac{s}{s^2 + 1/LC}$$

which on inversion gives

$$i = \frac{E_0}{L}\cos\left[\frac{t}{(LC)^{1/2}}\right]$$

EXERCISE 5

Determine the Laplace transforms of the following:

1 $A \delta(t)$ where A is a constant.
2 $8\delta(t + 3c)$ where c is a constant.
3 $F_0 \delta(t - 3)$ where F_0 is a constant.
4 $5 \delta(2t - 5)$.

Solve the following differential equations:

5 $\dfrac{d^2 x}{dt^2} + 16x = P_0 \delta(t)$

given that $x = 1$, $dx/dt = 0$ when $t = 0$ and P_0 is constant.

6 $\dfrac{d^2Y}{dt^2} + 9Y = \delta(t)$

given that $Y = 0$, $dY/dt = 2$ when $t = 0$.

7 $\dfrac{d^2Y}{dt^2} + 4Y = \delta(t - 1)$

given that $Y = 1$, $dY/dt = 0$ when $t = 0$.

8 $R\dfrac{dq}{dt} + \dfrac{q}{C} = E_0 + E_1\,\delta(t - 1)$

given that $q = 0$ when $t = 0$ and R, C, E_0 and E_1 are constants.

9 A mass is attached to the lower end of a vertical spring suspended from a fixed point. At time $t = t_0$ the mass is struck by an impulsive force in the upward direction of magnitude P_0. If the damping is assumed to be proportional to the velocity then the displacement x is given by the differential equation

$$m\frac{d^2x}{dt^2} + c\frac{dx}{dt} + kx = P_0\,\delta(t - t_0)$$

Solve this equation if $x = 0$, $dx/dt = 0$ at $t = 0$ assuming that $4km^2 - c^2 > 0$.

10 A light beam of length l rests horizontally on supports at each end and carries a load W at a point $\frac{1}{3}l$ from the left-hand end. Solve the equation

$$EI\frac{d^4y}{dx^4} = W\,\delta\!\left(x - \frac{l}{3}\right)$$

subject to the conditions $y = 0$ at $x = 0$ and at $x = l$ and at each free end d^2y/dx^2 (the bending moment) is zero.

11 A circuit is subjected to an infinite series of impulsive voltages of magnitude E_0 at $t = 0, 1, 2, \dots$. If the current in the circuit is given by

$$\frac{d^2i}{dt^2} + 4\frac{di}{dt} + 3i = E$$

determine i in terms of t if $i = 0$, $di/dt = 0$ when $t = 0$.

6

Transfer functions, block diagrams and stability

Transfer functions

In the study of equations representing physical systems, it is necessary to consider the relationships between the input and output variables of the system. Consider a system governed by the linear differential equation

$$a\,\frac{d^2x}{dt^2} + b\,\frac{dx}{dt} + cx = f(t) \qquad\qquad (6.1)$$

where a, b, c are constants and all the initial conditions are zero. In a stable system the roots of the auxiliary equation must have negative real parts so that the ratio b/a is positive.

Taking Laplace transforms of both sides produces the algebraic equation

$$(as^2 + bs + c)\bar{x}(s) = \bar{f}(s)$$

where s is the Laplace transform variable and $\bar{x}(s), \bar{f}(s)$ are the Laplace transforms of x and $f(t)$ respectively. The function $as^2 + bs + c$ is called the characteristic function and the equation

$$as^2 + bs + c = 0$$

is the characteristic equation.

Since the transformed equation is algebraic, then

$$\frac{\bar{x}(s)}{\bar{f}(s)} = \frac{1}{as^2 + bs + c}$$

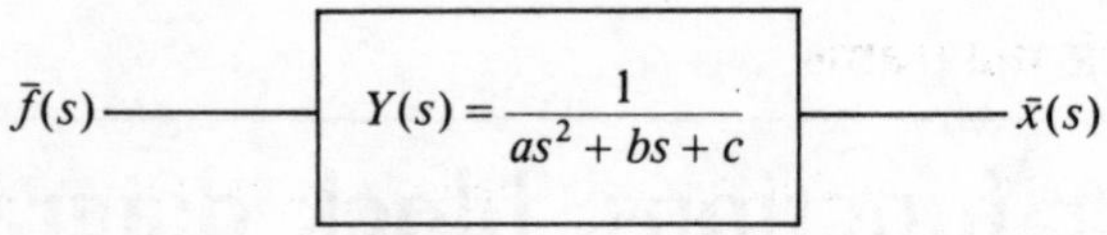

Figure 6.1 A block diagram

The function on the right-hand side is the *transfer function* relating $\bar{x}(s)$ and $\bar{f}(s)$ and is the ratio

$$\frac{\text{Laplace transform of output}}{\text{Laplace transform of input}}$$

The usual notation for a transfer function is $Y(s), G(s)$. A block diagram (Figure 6.1) can be used to represent such a relationship.

A transfer function can also be obtained by one of the following methods.

Trial solution method

Assume a particular integral $x = Ae^{j\omega t}$. By substitution into equation (6.1) we obtain

$$\frac{x}{f(t)} = \frac{1}{a(j\omega)^2 + b(j\omega) + c}$$

The function on the right-hand side is called the transfer function of the system and is the ratio of output to input.

This form will be encountered in the discussion of the Fourier transform.

D operator

If $D = d/dt$ and $D^2 = d^2/dt^2$ then equation (6.1) can be written

$$(aD^2 + bD + c)x = f(t)$$

so that

$$\frac{x}{f(t)} = \frac{1}{aD^2 + bD + c}$$

Note the similarity of the transfer function using $s, j\omega$ or D.

Simple block diagrams

A system governed by a differential equation can be represented diagrammatically by a block, or combination of blocks, each consisting of a transfer function $Y(s)$ with input $\bar{f}(s)$ and output $\bar{x}(s)$.

Consider the system of an amplifier, a d.c. motor and a load shown in Figure 6.2. To facilitate the writing of the equations the system can be subdivided into smaller parts as illustrated. The equivalent block diagram can then be obtained:

$$\frac{\bar{e}_2}{\bar{e}_1} = Y_1(s) \tag{6.2}$$

$$\frac{\bar{i}_f}{\bar{e}_2} = Y_2(s) \tag{6.3}$$

$$\frac{\bar{T}}{\bar{i}_f} = Y_3(s) \tag{6.4}$$

$$\frac{\bar{\theta}_o}{\bar{T}} = Y_4(s) \tag{6.5}$$

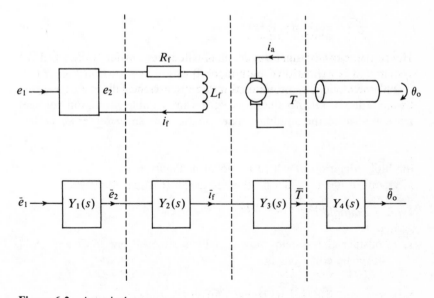

Figure 6.2 A typical system

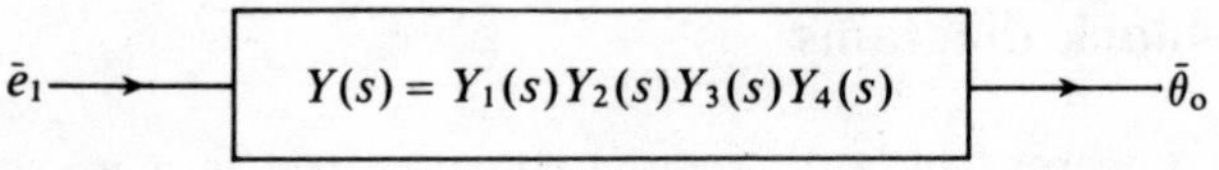

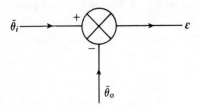

Figure 6.3

$\bar{\theta}_i \longrightarrow \overset{+}{\underset{-}{\bigotimes}} \longrightarrow \varepsilon$

$\bar{\theta}_o$

Figure 6.4 The error sensor

By substituting among these equations the intermediate variables can be eliminated and the relationship between the Laplace transform of the output ($\bar{\theta}_o$) and the Laplace transform of the input ($\bar{e}_1$) can be obtained.

$$\bar{\theta}_o = Y_4(s)\bar{T} = Y_4(s)Y_3(s)\bar{i}_f = \cdots$$

Therefore

$$\frac{\bar{\theta}_o}{\bar{e}_1} = Y_4(s)Y_3(s)Y_2(s)Y_1(s) \tag{6.6}$$

Hence the transfer functions of blocks in series are multiplied and the system can be represented by a block diagram as shown in Figure 6.3.

In practice it may be necessary to compare the output θ_o to a particular value or set of values, say θ_i, so that suitable correction voltages ε are applied to the amplifier. The equation for an *error sensor* is then

$$\bar{\varepsilon} = \bar{\theta}_i - \bar{\theta}_o \tag{6.7}$$

the block diagram of which is shown in Figure 6.4.

Typical examples

1. Consider the system shown in Figure 6.5. From (6.6) and (6.7) above the equations are

$$\frac{\bar{\theta}_o}{\bar{\varepsilon}} = Y_1(s)Y_2(s) \qquad \bar{\varepsilon} = \bar{\theta}_i - \bar{\theta}_o$$

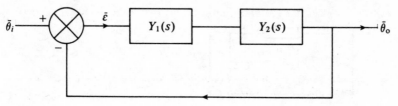

Figure 6.5

Substituting for $\bar{\varepsilon}$ we have

$$\bar{\theta}_o = Y_1(s)Y_2(s)\{\bar{\theta}_i - \bar{\theta}_o\}$$

$$\bar{\theta}_o\{1 + Y_1(s)Y_2(s)\} = Y_1(s)Y_2(s)\bar{\theta}_i$$

Therefore

$$\frac{\bar{\theta}_o}{\bar{\theta}_i} = \frac{Y_1(s)Y_2(s)}{1 + Y_1(s)Y_2(s)}$$

2. A system (Figure 6.6) may also have a transfer function in the feed-back loop. The equations are

$$\frac{\bar{\theta}_o}{\bar{\varepsilon}} = Y_1(s)Y_2(s) \qquad \frac{\bar{x}}{\bar{\theta}_o} = Y_3(s) \qquad \bar{\varepsilon} = \bar{\theta}_i - \bar{x}$$

By substitution the overall transfer function becomes

$$\frac{\bar{\theta}_o}{\bar{\theta}_i} = \frac{Y_1(s)Y_2(s)}{1 + Y_1(s)Y_2(s)Y_3(s)}$$

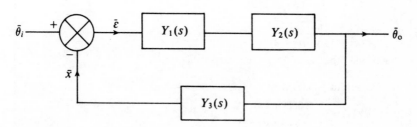

Figure 6.6

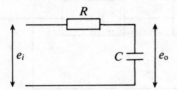

Figure 6.7 A CR network

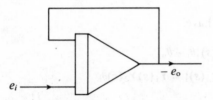

Figure 6.8 An operational amplifier

For purposes of analysis transfer functions can be represented and studied by means of CR networks (Figure 6.7)

$$\frac{\bar{e}_o}{\bar{e}_i} = \frac{1}{1 + CRs}$$

or operational amplifiers (Figure 6.8)

$$\frac{\bar{e}_o}{\bar{e}_i} = \frac{1}{s + 1}$$

EXAMPLE 6.1

Determine the steady state gain of the system whose transfer function is

$$\frac{100(1 + 2s)}{(s^2 + 8s + 6)(s + 3)}$$

when the input is $u(t)$, the unit step function.

If the output is $y(t)$ then

$$Y(s) = \frac{100(1 + 2s)}{(s^2 + 8s + 6)(s + 3)} \mathcal{L}\{u(t)\}$$

$$= \frac{100(1 + 2s)}{(s^2 + 8s + 6)(s + 3)} \frac{1}{s}$$

The steady state gain for unit input is defined by

$$\lim_{t \to \infty} y(t) = \lim_{s \to 0} sY(s) \text{ (by the final value theorem)}$$

$$= \lim_{s \to 0} \frac{100(1 + 2s)}{(s^2 + 8s + 6)(s + 3)}$$

$$= \frac{100}{18}$$

Steady state error

The prime reason for using feedback is to minimize the error between the actual system output and the desired system output. The magnitude of the steady state error, the value to which the error signal tends as the transient disturbance from any change of input dies away, is a measure of the accuracy of the system.

Consider the system defined in Figure 6.9. The equations are

$$\bar{\theta}_o = Y_1(s)\bar{\varepsilon} \qquad \bar{\varepsilon} = \bar{\theta}_i - Y_2(s)\bar{\theta}_o$$

and by eliminating $\bar{\theta}_o$ we obtain

$$\bar{\varepsilon} = \frac{\bar{\theta}_i}{1 + Y_1(s)Y_2(s)}$$

By definition

$$\text{steady state error} = \lim_{t \to \infty} e(t) = \lim_{s \to 0} s\bar{\varepsilon}$$

by the final value theorem. Therefore

$$\text{steady state error} = \lim_{s \to 0} \frac{s\bar{\theta}_i}{1 + Y_1(s)Y_2(s)}$$

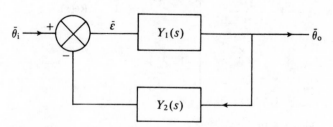

Figure 6.9

Poles, zeros and stability

Let us consider the general form of a transfer function $G(s)$ given by

$$G(s) = \frac{b_0 + b_1 s + b_2 s^2 + \cdots}{a_0 + a_1 s + a_2 s^2 + \cdots} = \frac{N(s)}{D(s)}$$

If $N(s)$ and $D(s)$ are of degree k and degree m respectively then they can be factorized into k factors and m factors respectively. The constants involved may be real or complex.

The roots of $N(s) = 0$ are called the *zeros* of the transfer function $G(s)$, whilst the roots of $D(s) = 0$ are called the *poles*. The power of each factor determines the *order* of the pole or zero. The stability and corresponding response of a system can be determined from the location of the poles.

EXAMPLE 6.2

$$G(s) = \frac{s(s-1)(s-0.5)}{(s+1)^2(s^2+s+1)}$$

The zeros are at $s = 0$, $s = 1$, $s = 0.5$. The poles are at $s = -1$ and $s = -1/2 \pm j\sqrt{3}/2$. Since $G(s) \to 0$ as $s \to +\infty$, $G(s)$ is also considered to have a zero at infinity.

Since s may be complex all finite values of poles and zeros can then be plotted on an Argand diagram (Figure 6.10) where $\times$ and $\bigcirc$ mark the poles and zeros respectively.

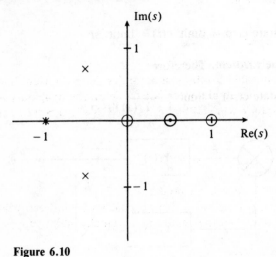

Figure 6.10

An output function defined by $Y(s)$ is a combination of two functions, i.e.

$$Y(s) = G(s)X(s)$$

where $G(s)$ is the transfer function of the system and $X(s)$ is dependent on the input to that system. Hence the poles of $Y(s)$ can arise from

1. the poles of $G(s)$ and
2. the poles of $X(s)$.

The portion due to the poles of $G(s)$ is defined as the *natural response* and that due to the poles of $X(s)$ the *forced response*. If the system has all its poles in the left-hand half of the s plane these values of s have negative real parts so that the natural response is a combination of negative exponential terms which tend to zero as $t \to \infty$, i.e. a transient response.

Stability considerations

By a stable system we mean a system that yields a finite output for every finite input. Stability considerations can be represented by the transfer function $G(s)$.

A system is *stable* if all the poles of $G(s)$ lie in the left-hand half of the s plane since again the response will contain only negative exponential terms, e.g.

$$G(s) = \frac{1}{s + a} \qquad a > 0$$

represents a stable system, pole at $s = -a$.

A system is unstable if one or more poles lie in the right-hand half of the s plane or if multiple-order pole(s) lie on the imaginary axis. In this case the response will contain terms such as e^{at}, $a > 0$ or $t\cos(\omega t)$. For example,

$$G(s) = \frac{1}{s - a} \qquad a > 0$$

represents an *unstable* system, pole at $s = a$.

If $G(s)$ defines a function which approaches a non-zero value or a bounded oscillation then the system is marginally stable, e.g.

$$G(s) = \frac{1}{s^2 + a^2}$$

The first-order complex conjugate pair $s = \pm ja$ defines an oscillatory system.

Note that the coefficients in the numerator which determine the zeros only affect the constants in the calculation of partial fractions and not the form of those fractions. The zeros therefore contribute to amplitude and phase but do not influence the form of the time function and hence stability.

Sinusoidal steady state response

If the driving function (input) is sinusoidal, e.g. $A \sin(\omega t)$ or $\text{Im}\{A \exp(j\omega t)\}$, then the steady state (particular integral) is also sinusoidal of the same frequency but with different amplitude and phase, i.e.

$$E \sin(\omega t) + F \cos(\omega t) = R \sin(\omega t + \phi)$$

where R and ϕ depend on ω and of course the constants of the system under consideration. This dependence is conveniently obtained by replacing s by $j\omega$

$$Y(j\omega) = G(j\omega)X(j\omega)$$

and discussion confined to the function $G(j\omega)$ which, in general, will be complex having magnitude $|G(j\omega)|$, argument $\beta(j\omega)$:

$G(j\omega)$	steady state transfer function
$\mid G(j\omega) \mid$	magnitude (amplitude) response
$\beta(j\omega)$	phase response

Hence given $G(s)$ we can evaluate $|G(j\omega)|$ and $\beta(j\omega)$ as functions of ω.

NOTE 1   Gain M is defined as $M = 20 \log_{10}|G(j\omega)|$ and a Bode plot is graphs of M and β against ω. Such information is helpful in determining the necessary additions to stabilize a system or improve its response.

NOTE 2   Output amplitude = system magnitude × input amplitude

since $|Y(j\omega)| = |G(j\omega)X(j\omega)| = |G(j\omega)||X(j\omega)|$

and

output phase = system phase + input phase

since

$$\arg[Y(j\omega)] = \arg[G(j\omega)X(j\omega)]$$

$$= \arg[G(j\omega)] + \arg[X(j\omega)]$$

(the multiplication of complex expressions in polar form).

EXAMPLE 6.3

Determine the magnitude response and phase response of

$$G(s) = \frac{1+s}{1+s+s^2}$$

$$G(s) = \frac{1+s}{1+s+s^2}$$

$$G(j\omega) = \frac{1+(j\omega)}{1+(j\omega)+(j\omega)^2}$$

$$= \frac{1+j\omega}{(1-\omega^2)+j\omega}$$

magnitude response $= |G(j\omega)|$

$$= \left|\frac{1+j\omega}{(1-\omega^2)+j\omega}\right|$$

$$= \frac{|1+j\omega|}{|(1-\omega^2)+j\omega|}$$

$$= \frac{(1+\omega^2)^{1/2}}{[(1-\omega^2)^2+\omega^2]^{1/2}}$$

phase response $\quad = \arg[G(j\omega)]$

$$= \arg\left[\frac{1+j\omega}{(1-\omega^2)+j\omega}\right]$$

$$= \arg(1+j\omega) - \arg[(1-\omega^2)+j\omega]$$

$$= \tan^{-1}(\omega) - \tan^{-1}\left(\frac{\omega}{1-\omega^2}\right)$$

EXERCISE 6

Determine the poles, zeros, magnitude response and phase response of the following functions of s:

1   $\dfrac{s+1}{s^3}$ 2 $\dfrac{9s}{s^2-16}$

3   $\dfrac{1}{2s+1}$ 4 $\dfrac{s+3}{s^2+1}$

5   $\dfrac{s}{4s^2+1}$ 6 $\dfrac{s}{s^2+2s+2}$

7

Convolution

The concept of convolution is inherent in almost every field of the physical sciences and engineering. For example, in mechanics it is known as the superposition or Duhamel integral; in systems theory it plays a crucial role as the impulse response integral, and in optics as the point spread or smearing function.

Let $f(t)$ and $g(t)$ be defined for $t > 0$ and piecewise continuous. Then the *convolution* of f and g, written $f*g$, is defined by

$$f*g = \int_0^t f(t-u)g(u)\,du \qquad 0 \leqslant t < \infty \qquad (7.1)$$

which is another function of t defined for positive t.

EXAMPLE 7.1

Determine the convolution of $f(t)$ and $g(t)$ if $f(t) = e^t$, $g(t) = t^2$.

$$f*g = \int_0^t u^2 e^{t-u}\,du = 2e^t - t^2 - 2t - 2$$

obtained by integrating by parts twice.

EXAMPLE 7.2

Determine $f*g$ if

$$g(t) = \begin{cases} 0 & 0 \leqslant t < 3 \\ 2 & t \geqslant 3 \end{cases}$$

and $f(t) = \sin(t)$

▶

$$f(t - u) = \sin(t - u) \qquad g(u) = 2 \qquad u \geqslant 3$$

so that

$$f * g = \int_3^t 2 \sin(t - u) \, du \qquad \text{for } 3 \leqslant t$$

Hence

$$f * g = \begin{cases} 0 & 0 \leqslant t < 3 \\ 2 - 2\cos(t - 3) & 3 \leqslant t \end{cases}$$

which may be written $[2 - 2\cos(t - 3)] H(t - 3)$.

The convolution has several special properties of which the most important is

$$\mathcal{L}\{f * g\} = \mathcal{L}\{f\}\mathcal{L}\{g\} = F(s)G(s)$$

(Appendix 2, number 10). This result is particularly useful in finding inverse transforms, i.e.

$$\mathcal{L}^{-1}\{F(s)G(s)\} = f * g$$

Another major use is claimed to be the evaluation of convolution integrals by transforming to functions of s and then inverting the product by some other method.

Other properties are easily obtained from the definition:

- $f * g = g * f$
- $f * cg = (cf) * g = c(f * g)$ where c is a constant
- $f * (g_1 + g_2) = f * g_1 + f * g_2$
- $f_1 * (f_2 * f_3) = (f_1 * f_2) * f_3$
- $\delta(t) * f = f * \delta(t) = f$, $\delta(t - a) * f = f(t - a)$, $a > 0$, by evaluating the convolution integral.

EXAMPLE 7.3

Determine

$$\mathcal{L}^{-1}\left\{ e^{-as} \frac{1}{(s + 1)^2} \right\}$$

$$\mathcal{L}^{-1}\{e^{-as}\} = \delta(t - a) \qquad \mathcal{L}^{-1}\left\{ \frac{1}{(s + 1)^2} \right\} = t$$

$$\mathscr{L}^{-1}\left\{e^{-as}\frac{1}{(s+1)^2}\right\} = \int_0^t \delta(u-a)(t-u)\,\mathrm{d}u = \begin{cases} 0 & t < a \\ t-a & t \geqslant a \end{cases}$$

which may be written $(t-a)H(t-a)$.

EXAMPLE 7.4

Using Laplace transforms and the convolution integral solve the differential equation

$$\frac{\mathrm{d}^2 i}{\mathrm{d}t^2} + \omega^2 i = f(t)$$

where $i = \mathrm{d}i/\mathrm{d}t = 0$ when $t = 0$ and $f(t)$ is an unspecified function.

Transforming the equation gives

$$(s^2\bar{i} - si_0 - i_0') + \omega^2\bar{i} = \mathscr{L}\{f(t)\} = F(s)$$

Hence

$$\bar{i} = \frac{F(s)}{s^2+\omega^2} = \frac{1}{s^2+\omega^2}F(s)$$

Inverting using Appendix 2, number 10, we obtain

$$i = \int_0^t f(u)\,\frac{1}{\omega}\sin[\omega(t-u)]\,\mathrm{d}u$$

or

$$\int_0^t f(t-u)\,\frac{1}{\omega}\sin(\omega u)\,\mathrm{d}u$$

since $f*g = g*f$. Hence, given $f(t)$, the current may be found by evaluating one of these integrals analytically, graphically or numerically.

EXAMPLE 7.5

Determine the convolution of the functions

$$h(t) = \begin{cases} t^2 & t \geqslant 0 \\ 0 & t < 0 \end{cases} \qquad x(t) = \begin{cases} 1 & 0 \leqslant t < 1 \\ 0 & \text{elsewhere} \end{cases}$$

Method 1: Analytic evaluation of the convolution integral

$$h(t) = t^2 \qquad t \geqslant 0$$

$$x(t) = 1 - H(t-1)$$

where $H(t)$ denotes the step function, the alternative notation being used to avoid any confusion with the variable of integration.

$$h * x = \int_0^t [1 - H(u - 1)](t - u)^2 \, du$$

$$= \int_0^t (t - u)^2 \, du - \int_0^t H(u - 1)(t - u)^2 \, du$$

$$= \begin{cases} [-\tfrac{1}{3}(t - u)^3]_0^t - \int_1^t (t - u)^2 \, du & t \geqslant 1 \\ \\ [-\tfrac{1}{3}(t - u)^3]_0^t & t < 1 \end{cases}$$

$$= \tfrac{1}{3} t^3 + [\tfrac{1}{3}(t - u)^3]\{H(t - 1)$$

$$= \tfrac{1}{3} t^3 - [\tfrac{1}{3}(t - 1)^3] H(t - 1)$$

Method 2: Graphical interpretation and evaluation of the convolution integral. Let

$$y = h * x = \int_0^t x(u)h(t - u) \, du \quad t \geqslant 0$$

or

$$\int_0^t h(u)x(t - u) \, du \quad t \geqslant 0$$

Figures 7.1(a) and 7.1(b) show graphs of $h(u)$ and $x(u)$. The identification of points A and B emphasizes the operations of folding (Figure 7.1(c), graph of $x(-u)$) and shift (Figure 7.1(d), graph of $x(t - u)$). Figure 7.1(e) shows the graph of $h(u)x(t - u)$ for $0 \leqslant t < 1$. Hence

$$y = \int_0^t 1 \times u^2 \, du = \tfrac{1}{3} t^3 \qquad 0 \leqslant t < 1 \tag{7.2}$$

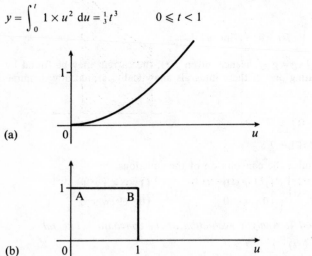

(a)

(b)

Figure 7.1 (a) Graph of $h(u)$; (b) graph of $x(u)$; (c) graph of $x(-u)$ (folding); (d) graph of $x(t - u)$ (shift); (e) graph of $h(u)x(t - u)$, $0 \leqslant t < 1$; (f) graph of $h(u)x(t - u)$, $t \geqslant 1$

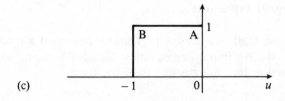

(c)

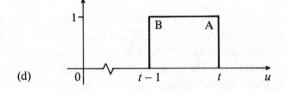

(d)

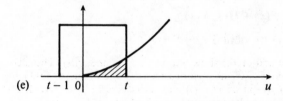

(e)

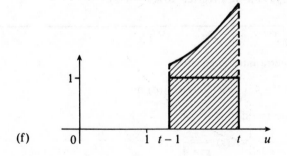

(f)

The graph of $h(u)x(t-u)$ for $t > 1$ (Figure 7.1(f)) gives

$$y = \int_{t-1}^{t} 1 \times u^2 \, du = \tfrac{1}{3}t^3 - \tfrac{1}{3}(t-1)^3 \qquad t \geqslant 1 \qquad (7.3)$$

Hence combining equations (7.2) and (7.3) the result is

$$y = \tfrac{1}{3}t^3 - \tfrac{1}{3}(t-1)^3 H(t-1)$$

Some texts refer to the convolution integral as the *Faltung integral* from the German word for folding.

Volterra integral equation

In 1931 the Italian mathematician Vito Volterra published a model of population growth. An important equation called a Volterra integral equation occurred in this analysis and is of the form

$$x(t) = f(t) + \int_0^t h(t-u)x(u)\ \mathrm{d}u \qquad (7.4)$$

where $x(t)$ is an unknown function and $f(t), h(t)$ are known functions. Subsequently many problems in thermodynamics, electrical systems, nuclear reactor theory and chemotherapy have been modeled with this type of equation. Note that the integral in equation (7.4) is a convolution integral so that taking Laplace transforms and using number 10 of Appendix 2 gives

$$X(s) = F(s) + H(s)X(s)$$

Solving for $X(s)$ we obtain

$$X(s) = \frac{F(s)}{1 - H(s)}$$

which, in theory, can be inverted to give $x(t)$.

EXAMPLE 7.6

Solve the Volterra integral equation

$$x(t) = \mathrm{e}^{-t} - 4 \int_0^t \cos[2(t-u)x](u)\ \mathrm{d}u$$

Taking Laplace transforms

$$X(s) = \frac{1}{s+1} - 4X(s)\frac{s}{s^2+4}$$

$$X(s)\left(1 + \frac{4s}{s^2+4}\right) = \frac{1}{s+1}$$

$$X(s)\frac{(s+2)^2}{s^2+4} = \frac{1}{s+1}$$

$$X(s) = \frac{s^2+4}{(s+1)}$$

$$= \frac{5}{s+1} - \frac{4}{(s+2)} + \frac{8}{(s+2)^2}$$

by partial fractions. Inverting we obtain $x(t) = 5\mathrm{e}^{-t} - 4\mathrm{e}^{-2t} - 8t\mathrm{e}^{-2t}$.

EXERCISE 7

1 Determine the convolution $f*g$ and verify that $\mathscr{L}\{f*g\} = \mathscr{L}\{f\}\mathscr{L}\{g\}$ for each of the following:

(a) $f(t) = 5, \ t \geqslant 0$ $g(t) = 2t, \ t \geqslant 0$

(b) $f(t) = t - 1, \ t \geqslant 0$ $g(t) = 3t^2, \ t \geqslant 0$

(c) $f(t) = \cos(t), \ t \geqslant 0$ $g(t) = \sin(3t), \ t \geqslant 0$

(d) $f(t) = e^{-t}, \ t \geqslant 0$ $g(t) = t + 2, \ t \geqslant 0$

(e) $f(t) = \begin{cases} t & 0 < t \leqslant 1 \\ 0 & t > 1 \end{cases}$ $g(t) = e^{-t}, \ t \geqslant 0$

(f) $f(t) = \begin{cases} 4 & 0 < t \leqslant 2 \\ 0 & t > 2 \end{cases}$ $g(t) = \sin(2t), \ t \geqslant 0$

2 Write the inverse Laplace transforms of the following functions as a convolution:

(a) $\dfrac{3}{(s+3)(s+1)}$ (b) $\dfrac{5s}{(s^2+9)(s-4)}$

(c) $\dfrac{e^{-3s}}{s(s^2+16)}$ (d) $\dfrac{e^{-2s}}{s+3}$

(e) $\left(\dfrac{2e^{-s}}{s} - \dfrac{3e^{-2s}}{s}\right) \dfrac{s}{s^2+9}$

3 Evaluate the convolutions

(a) $\delta(t)*\cos(4t)$ (b) $\delta(t-2)*e^{-2t}u(t)$

4 Solve the following differential equations giving the result as a convolution integral:

(a) $\dfrac{d^2y}{dt^2} + 2a\dfrac{dy}{dt} + a^2y = f(t)$

given that $y = 0$, $dy/dt = 0$ when $t = 0$, a is constant and $f(t)$ is an unspecified function of t.

(b) $\dfrac{d^2i}{dt^2} + (a+b)\dfrac{di}{dt} + abi = f(t)$

given that $i = 0$, $di/dt = 0$ when $t = 0$, and a, b are constants.

(c) $\dfrac{d^2x}{dt^2} + 5\dfrac{dx}{dt} + 6x = \dfrac{1}{1+t}$

given that $x = 0$, $dx/dt = 2$ when $t = 0$. *Do not attempt to determine* $\mathscr{L}\{1/(1+t)\}$.

5 The actual angle θ in a control mechanism is given by the differential equation

$$\frac{d^2\theta}{dt^2} = -K^2(\theta - \phi) \qquad K \text{ constant}$$

If ϕ is a function of t and $\theta = 0$, $d\theta/dt = 0$ when $t = 0$, determine θ as a convolution integral in terms of ϕ.

6 Solve the following equations:

(a) $x(t) = t + 2 \displaystyle\int_0^t \sin(t - u)x(u) \, du$

(b) $x(t) = 3t^2 + \dfrac{1}{6} \displaystyle\int_0^t (t - u)^3 x(u) \, du$

8

Sampled-data control systems and the z transform

In certain types of feedback control systems the signal appears as a set of sequence values. This signal might be a discrete-time signal in its own right or derived from a continuous-time function by a sampling operation. In either case the signal can be considered as a train of impulses equally spaced at discrete moments of time. Such systems are known as sampled-data control systems or discrete-time systems. The sampling operation introduces a time delay and a loss of information between samples which may lead to instability in the control system. The analysis of such systems can be undertaken using difference equations and the z transform in the same way that continuous-time systems were represented by differential equations and the Laplace transform. The impact of the digital computer, the transmission of digitized data with greater accuracy than continuous data and the use of time-shared communication have all contributed to the interest in sampled-data theory.

Consider a continuous function $f(t)$. If this function is sampled at equal intervals of time T, the sampled function is a sequence of numbers

$$f(0), f(T), f(2T), f(3T), ..., f(nT), ...$$

This series of numbers gives a limited description of $f(t)$. Its values at other times can only be approximated by means of interpolation and extrapolation (Figure 8.1). The sampled information will be designated $f^*(t)$, the graph of which will be a bar chart or simply a set of impulses of appropriate magnitude at $t = 0, T, 2T, 3T, ...$. Data predictors can be used to construct $f(t)$ from the information $f^*(t)$ (Shinners, 1964).

Note that discontinuous functions must be defined precisely at the

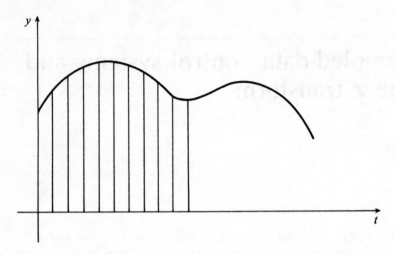

Figure 8.1 Sampling of $f(t)$

points of discontinuity, e.g. $u(t - k)$ at $t = k$ is defined, in Chapter 4, to be equal to $+1$.

The sampled function $f^*(t)$ may be represented, as in Chapter 5, property 7, by

$$f^*(t) = f(0)\ \delta(t) + f(T)\ \delta(t - T) + f(2T)\ \delta(t - 2T) + \cdots$$

$$= \sum_{n=0}^{\infty} f(nT)\ \delta(t - nT)$$

It is assumed that the sampling process is ideal and that no electronic switching time is considered.

The Laplace transform of this function can be written as

$$F^*(s) = \mathscr{L}\{f^*(t)\}$$

$$= \int_0^{\infty} \{f(0)\ \delta(t) + f(T)\ \delta(t - T) + f(2T)\ \delta(t - 2T)\ldots\}e^{-st}\ dt$$

$$= \qquad f(0) \qquad + f(T)e^{-sT} \qquad + f(2T)e^{-2sT}\ldots$$

$$= \sum_{n=0}^{\infty} f(nT)e^{-nTs}$$

If we now define a new variable z such that

$$z = e^{sT}$$

or

$$s = \frac{1}{T} \ln(z)$$

and denote $F^*(s)$ by $F(z)$ then

$$F(z) = \sum_{n=0}^{\infty} f(nT)z^{-n} = f(0) + f(T)z^{-1} + f(2T)z^{-2} + \cdots \quad \textbf{(8.1)}$$

which is a power series in z^{-1} or $1/z$. $F(z)$ is called the z transform of the sequence or discrete-time signal $f(nT)$ and the relationship is written as

$$F(z) = \mathcal{F}\{f(nT)\}$$

In general, if any continuous function possesses a Laplace transform then the corresponding sampled function has a z transform.

EXAMPLE 8.1

Determine the z transform of the sampled function $f(nT)$ if $f(t)$ is the unit step function $u(t)$.

Sampling this function gives $f(nT) = 1$, $n \geqslant 0$.

$$F(z) = \sum_{n=0}^{\infty} 1z^{-n} = 1 + z^{-1} + z^{-2} + z^{-3}\ldots$$

from equation (8.1). Hence

$$F(z) = \frac{1}{1 - z^{-1}} = \frac{z}{z - 1}$$

(this is the sum of a geometric progression that is convergent for $|z| > 1$). Therefore

$$\mathcal{F}\{u(nT)\} = \frac{z}{z - 1}$$

EXAMPLE 8.2

Determine the z transform of the sampled function $f(nT)$ if $f(t)$ is the ramp function $f(t) = t$.

Sampling gives $f(nT) = 0, T, 2T, 3T, \ldots$. From equation (8.1)

$$F(z) = Tz^{-1} + 2Tz^{-2} + 3Tz^{-3} + \cdots = \frac{T}{z}(1 - z^{-1})^{-2} = \frac{Tz}{(z - 1)^2}$$

which is convergent for $|z| > 1$. Therefore

$$\mathscr{J}\{nT\} = \frac{Tz}{(z-1)^2}$$

EXAMPLE 8.3

Determine the z transform of the sampled function $f(n)$ if $f(t) = e^{-at}$ where a is a constant.

Sampling gives $f(n) = e^{-an}$.

$$F(z) = 1 + e^{-a}z^{-1} + e^{-2a}z^{-2} + e^{-3a}z^{-3} + \cdots$$

$$= \frac{1}{1 - e^{-a}z^{-1}} = \frac{z}{z - e^{-a}}$$

which is convergent for $|z| > e^{-a}$. Therefore

$$\mathscr{J}\{e^{-an}\} = \frac{z}{z - e^{-a}}$$

In the same way that a table of Laplace transforms is derived from basic definitions, so also is a table of z transforms (Appendix 3). Note the following.

1. z may be complex so that conditions for convergence of the infinite series involve the modulus.
2. The z transform is a unique transformation.
3. It can be shown that the z transform is independent of T so that $f(nT)$ can be replaced by $f(n)$, the preferred notation. The sampling interval T is simply a scale factor.

Some important theorems

These results can be proved from the definition (Doetsch, 1971; Karni and Byatt, 1982).

1. Addition:

$$\mathscr{J}\{f_1(n) \pm f_2(n)\} = \mathscr{J}\{f_1(n)\} \pm \mathscr{J}\{f_2(n)\}$$

2. Multiplication by a constant:

$$\mathcal{F}\{af(n)\} = a\mathcal{F}\{f(n)\}$$

3. Translation – shift theorem:

$$\mathcal{F}\{f(n-m)u(n-m)\} = z^{-m}\mathcal{F}\{f(n)\}$$

$$\mathcal{F}\{f(n+m)\} = z^{m}\left[\mathcal{F}\{f(n)\} - \sum_{k=0}^{m-1} f(k)z^{-k}\right]$$

These are shown in the table of standard forms (23 and 19–22 respectively).

4. Initial value theorem:

$$\underset{t\to 0}{\text{limit}}\ [f^*(t)] = \underset{z\to\infty}{\text{limit}}\ [F(z)]$$

5. Final value theorem:

$$\underset{t\to\infty}{\text{limit}}\ [f^*(t)] = \underset{z\to 1}{\text{limit}}\ \left[\frac{z-1}{z}\ F(z)\right]$$

6. Periodic sequence theorem: if $f(n) = f(0), f(1), f(2,), ..., f(N-1)$ and the sequence of N values is repeated periodically then

$$F(z) = \frac{F_1(z)}{1 - z^{-N}}$$

where $F(z)$ is the z transform of the periodic function and $F_1(z)$ is the z transform of the sequence $f(0), ..., f(N-1)$. For example if $f(n) = 1, -1, -1, 1$, and these four values are repeated periodically then

$$F_1(z) = 1 - z^{-1} - z^{-2} + z^{-3}$$

and

$$F(z) = \frac{F_1(z)}{1 - z^{-4}}$$

$$= \frac{z(z^3 - z^2 - z + 1)}{z^4 - 1}$$

$$= \frac{z(z-1)}{z^2 + 1}$$

EXAMPLE 8.4

Using the table of standard forms determine the z transforms of the following sampled functions:

(a) $f(n) = n$

(b) $f(n) = Ke^{-an} - 2\sin(bn)$ if a, b and K are constants.

(c) $f(n) = e^{-2(n-2)}u(n-2)$

(d) $f(n) = 2^n(n^2 - n)$

(a) $f(n) = n$:

$$F(z) = \frac{z}{(z-1)^2}$$

using number 4.

(b) $f(n) = Ke^{-an} - 2\sin(bn)$:

$$F(z) = \frac{Kz}{z - e^{-a}} - \frac{2z\sin(b)}{z^2 - 2z\cos(b) + 1}$$

using numbers 6 and 8.

(c) $f(n) = e^{-2(n-2)}u(n-2)$:

$$F(z) = z^{-2}\mathscr{Z}\{e^{-2n}\}$$

$$= z^{-2}\frac{z}{z - e^{-2}}$$

using numbers 23 and 6.

(d) $f(n) = 2^n(n^2 - n)$. From number 27 it is necessary first to determine the transform of $n^2 - n$:

$$\mathscr{Z}\{n^2 - n\} = \frac{z(z+1)}{(z-1)^3} - \frac{z}{(z-1)^2} = \frac{2z}{(z-1)^3}$$

The required transform is

$$\frac{2(z/2)}{[(z/2) - 1]^3} = \frac{8z}{(z-2)^3}$$

EXERCISE 8

Determine the z transform of each of the following:

1 n

2 n^2

3 $(-\frac{2}{3})^n$

4 $1 - (-5)^n$

5 ne^{-4n}

6 $2 + 3e^{-2n}$

7 $\cos(n\frac{\pi}{2})$

8 $e^{-an}\sin(\omega n)$

9 $3\sin(2n) - 4\cos(3n)$

10 $3^n\sin(2n)$

11 $u(n) - 4u(n-1)$

12 $[1 + (-1)^n]e^{-nT}$

13 $u(n-k)$

14 $\cosh(an)$

15 $2n - u(n) + 3e^{-n}$

16 $2n - 3(n-1)u(n-1)$

17 $u(n) - e^{-n} - u(n-1) + e^{-(n-1)}u(n-1).$

a, ω, T and k are constants.

9
The inverse z transform

The z transform $F(z)$ of a given sequence $f(n)$ is unique. Hence the z transform and its inverse form a transform pair:

$$F(z) = \mathcal{F}\{f(n)\}$$

$$f(n) = \mathcal{F}^{-1}\{F(z)\}$$

We now consider the problem of obtaining $f(n)$ from a given $F(z)$. There are three different methods: the Power series method; the partial fraction expansion method; the residue method. The partial fraction method is restricted to the case where $F(z)$ is in the form of a rational function, while the other methods apply to a larger class of functions.

Power series method

Using the method for dividing two polynomials $F(z)$ is expressed in the form of a power series to obtain

$$F(z) = a_0 + a_1 z^{-1} + a_2 z^{-2} + a_3 z^{-3} + \cdots$$

In Chapter 8 the z transform was defined as

$$F(z) = \sum_{n=0}^{\infty} f(n)z^{-n} = f(0) + f(1)z^{-1} + f(2)z^{-2} + f(3)z^{-3} + \cdots$$

A comparison of these series shows that there is a one-to-one correspondence between the coefficients a_n and the required sequence values $f(n)$.

Hence

$$f(n) = a_n \qquad n \geqslant 0$$

EXAMPLE 9.1

Determine the inverse z transform of

$$F(z) = \frac{z^2 + z}{z^3 - 3z^2 + 3z - 1}$$

$$
\begin{array}{r}
z^{-1} + 4z^{-2} + 9z^{-3} + 16z^{-4}\ldots \\
z^3 - 3z^2 + 3z - 1 \enclose{longdiv}{z^2 + z} \\
\underline{z^2 - 3z + 3 - z^{-1}} \\
4z - 3 + z^{-1} \\
\underline{4z - 12 + 12z^{-1} - 4z^{-2}} \\
9 - 11z^{-1} + 4z^{-2} \\
\underline{9 - 27z^{-1} + 27z^{-2} - 9z^{-3}} \\
16z^{-1} - 23z^{-2} + 9z^{-3}
\end{array}
$$

Hence

$$F(z) = z^{-1} + 4z^{-2} + 9z^{-3} + 16z^{-4}\ldots$$

so that $a_0 = 0$, $a_1 = 1$, $a_2 = 4$, $a_3 = 9$, $a_4 = 16, \ldots$, and $f(0) = 0$, $f(1) = 1$, $f(2) = 4$, $f(3) = 9$, $f(4) = 16, \ldots$, which suggests that $f(n)$ *may be equal to n^2*.

A disadvantage of this method is that, in general, only the sequence values are obtained and additional techniques are necessary to obtain a closed form version of $f(n)$.

Other methods can be used to derive the series expansion, e.g. the binomial theorem can be used to expand $z/(z + 1)$ in a series of negative powers of z:

$$\frac{z}{z + 1} = (1 + z^{-1})^{-1} = 1 + z^{-1} + \cdots$$

Partial fraction method

The approach here parallels that used to obtain the inverse Laplace transform of $F(s)$. In this case, however, we work with $F(z)/z$ instead of $F(s)$. Why $F(z)/z$ and not $F(z)$? Examining our standard forms we note that functions of z occur in the form

$$\frac{z}{z-1}, \frac{z}{(z-1)^2}, \frac{z}{z-e^{-a}}$$

and not in the normal partial fraction form

$$\frac{1}{z-1}, \frac{1}{(z-1)^2}, \frac{1}{z-e^{-a}}$$

EXAMPLE 9.2

Determine $f(n)$ if

$$F(z) = \frac{4z}{3z^2 - 2z - 1}$$

$$\frac{F(z)}{z} = \frac{4}{3z^2 - 2z - 1} = \frac{4}{(3z+1)(z-1)} = \frac{A}{z-1} + \frac{B}{3z+1}$$

in partial fractions. Thus

$$4 = A(3z+1) + B(z-1)$$

Put $z = 1$ to obtain $A = 1$ and put $z = -1/3$ to obtain $B = -3$. Hence

$$F(z) = \frac{z}{z-1} - \frac{3z}{3z+1} = \frac{z}{z-1} - \frac{z}{z+1/3}$$

From the standard forms 3 and 12 in Appendix 3

$$\mathscr{Z}^{-1}\left\{\frac{z}{z-1}\right\} = 1 \qquad \mathscr{Z}^{-1}\left\{\frac{z}{z+1/3}\right\} = \left(-\frac{1}{3}\right)^n$$

Hence

$$f(n) = 1 - (-3)^{-n} \qquad n \geqslant 0$$

Note that the sampled function $f^*(t)$ becomes

$$f^*(t) = \sum_{n=0}^{\infty} [1 - (-3)^{-n}] \, \delta(t-n)$$

EXAMPLE 9.3

Determine $f(n)$ if

$$F(z) = \frac{z}{(z+3)(z+2)^2}$$

$$\frac{F(z)}{z} = \frac{1}{(z+3)(z+2)^2} = \frac{A}{z+3} + \frac{B}{z+2} + \frac{C}{(z+2)^2}$$

for which $A = 1$, $B = -1$, $C = 1$. Therefore

$$F(z) = \frac{z}{z+3} - \frac{z}{z+2} + \frac{z}{(z+2)^2}$$

Inverting using 12 and 13 in Appendix 3 gives

$$f(n) = (-3)^n - (-2)^n - \tfrac{1}{2}n(-2)^n$$
$$= (-3)^n - (-2)^n + n(-2)^{n-1} \qquad n \geqslant 0$$

Sampled values of $f(n)$ can be obtained by calculations for $n = 0, 1, 2, \ldots$ or by the expansion of $F(z)$ in powers of z^{-n} using the binomial theorem.

EXAMPLE 9.4

Determine the inverse z transform of

$$F(z) = \frac{z[z-k]}{z^2 - 3z + 1}$$

Note that the denominators of 14 and 15 in Appendix 3 can be made to fit $z^2 - 3z + 1$ by choosing

$$2\cosh(\omega_0) = 3$$

i.e.

$$\cosh(\omega_0) = 3/2$$

and since

$$\sinh(\omega_0) = [\cosh^2(\omega_0) - 1]^{1/2}$$

$$\sinh(\omega_0) = \frac{\sqrt{5}}{2}$$

so that 14 and 15 read

$$\frac{\sqrt{5}}{2} \frac{z}{z^2 - 3z + 1} \qquad \frac{z(z - \frac{3}{2})}{z^2 - 3z + 1}$$

Now

$$F(z) = \frac{z[z - \frac{3}{2}] + z[\frac{3}{2} - k]}{z^2 - 3z + 1} = \frac{z[z - \frac{3}{2}]}{z^2 - 3z + 1} + \frac{z[\frac{3}{2} - k]}{z^2 - 3z + 1}$$

Therefore

$$f(n) = \cosh(\omega_0 n) + \frac{2}{\sqrt{5}}(\tfrac{3}{2} - k)\sinh(\omega_0 n)$$

where $\omega_0 = \cosh^{-1}(\tfrac{3}{2})$.

Residue method

From the definition of the z transform of a given sequence $f(n)$ it can be shown (Aseltine, 1958; Doetsch, 1971; Karni and Byatt, 1982) that

$$f(n) = \frac{1}{2\pi j} \oint_C z^{n-1} F(z) \, dz \qquad n \geqslant 0$$

where C is any simple closed curve enclosing $|z| = R$, $|z| > R$ being the region of convergence.

This integral is known as a *contour* integral and can be evaluated using Cauchy's residue theorem. This theorem states that the value of the integral is the sum of the residues of $z^{n-1} F(z)$ corresponding to the poles of the function that lie inside a simple closed curve C that encloses $|z| = R$.

Pole

The function $G(z) = z^{n-1} F(z)$ has a pole of order n at $z = z_1$ if

$$\lim_{z \to z_1} G(z) = \infty$$

and the value of $(z - z_1)^n G(z)$ at $z = z_1$ is finite and non-zero. For example

$$G(z) = \frac{z + 2}{z(z + 3)^2}$$

has a first-order pole (referred to as a simple pole) at $z = 0$ and a pole of second order at $z = -3$.

Residue

Associated with each pole of a function is a number called the residue of the function at the pole, which can be calculated from the formula

$$R_{z=a} = \frac{d^{m-1}}{dz^{m-1}} \frac{(z - a)^m G(z)}{(m - 1)!} \qquad m \geqslant 1$$

evaluated at $z = a$ and where m is the order of the pole and R the residue.

EXAMPLE 9.5

Determine the residues of

$$G(z) = \frac{z^2}{(z+3)(z-1)^2}$$

There is a pole of order 1 at $z = -3$:

$$R_{-3} = (z+3)G(z) = \frac{z^2}{(z-1)^2}$$

For $z = -3$ we obtain

$$R_{-3} = 9/16$$

There is a pole of order 2 at $z = 1$:

$$R_1 = \frac{d}{dz}\left[(z-1)^2 G(z)\right] = \frac{d}{dz}\left[\frac{z^2}{(z+3)}\right]$$

$$= \frac{(z+3)2z - z^2}{(z+3)^2}$$

At $z = 1$

$$R_1 = 7/16$$

Note that the residue at a pole corresponds to the coefficient of the *first*-order term in a partial fraction expansion.

$$\frac{z^2}{(z+3)(z-1)^2} = \frac{A}{z+3} + \frac{B}{z-1} + \frac{C}{(z-1)^2}$$

which requires $A = 9/16$, $B = 7/16$, $C = 1/4$, so that $R_{-3} = 9/16$ and $R_1 = 7/16$.

EXAMPLE 9.6

Determine the inverse z transform of

$$F(z) = \frac{z^2}{(z+3)^2}$$

$$G(z) = z^{n-1}F(z) = \frac{z^{n+1}}{(z+3)^2}$$

There is a pole of second order at $z = -3$:

$$R_{-3} = \frac{d}{dz}\left[(z+3)^2 G(z)\right] = \frac{d}{dz}\left[z^{n+1}\right]$$

$$= (n+1)z^n$$

▶

At $z = -3$

$$R_{-3} = (n+1)(-3)^n$$
$$f(n) = (n+1)(-3)^n$$

EXAMPLE 9.7

Determine the inverse z transform of

$$F(z) = \frac{1}{(z+1)(z+2)}$$

Consider

$$G(z) = z^{n-1}F(z) = \frac{z^{n-1}}{(z+1)(z+2)}$$

which has a simple pole at $z = 0$ when $n = 0$ but *not* at $z = 0$ for $n \geqslant 1$. Thus the cases $n = 0$ and $n \geqslant 1$ must be considered separately.

For $n = 0$

$$G(z) = \frac{1}{z(z+1)(z+2)}$$

$$R_{z=0} = \frac{1}{(1)(2)} = \frac{1}{2}$$

$$R_{z=-1} = \frac{1}{(1)(-1)} = -1$$

$$R_{z=-2} = \frac{1}{(-2)(-1)} = \frac{1}{2}$$

$$f(0) = \Sigma \ R = \tfrac{1}{2} - 1 + \tfrac{1}{2} = 0$$

For $n \geqslant 1$

$$G(z) = \frac{z^{n-1}}{(z+1)(z+2)}$$

$$R_{z=-1} = \frac{(-1)^{n-1}}{(1)} = (-1)^{n-1}$$

$$R_{z=-2} = \frac{(-2)^{n-1}}{(-1)} = -(-2)^{n-1}$$

Hence

$$f(n) = (-1)^{n-1} - (-2)^{n-1} \qquad n \geqslant 1$$

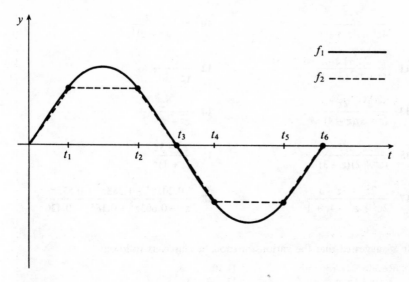

Figure 9.1

Note that the function $f(n)$ or $f(nT)$ obtained using either of these methods only describes the function *at the sampling instants*. There is no information available between sampling instants. For example, both of the functions f_1 and f_2 in Figure 9.1 pass through the given data points at t_1, t_2, t_3, t_4, t_5 and t_6.

EXERCISE 9

Determine the inverse z transforms of the following.

1 $\dfrac{z}{(z-3)^2}$ **2** $\dfrac{z}{z^2+1}$

3 $\dfrac{z^2}{z^2+1}$ **4** $\dfrac{z}{z^2-z+1}$

5 $\dfrac{z^2}{z^2-z+1}$ **6** $\dfrac{z}{z^2-4z+1}$

7 $\dfrac{z^2}{z^2-4z+1}$ **8** $\dfrac{4z}{4z^2-2z\sqrt{3}+1}$

9 $\dfrac{4z^2}{4z^2 - 2z\sqrt{3} + 1}$ 10 $z^{-3}\dfrac{2z}{(z+2)^2}$

11 $\dfrac{2z^2 - z}{(z+1)^2(z-2)}$ 12 $\dfrac{2z^3}{(z-2)^3}$

13 $\dfrac{z^2 + 2z}{(z+3)(z-4)}$ 14 $\dfrac{2z}{z^2 + z - 2}$

15 $\dfrac{3z^2 + 2z}{(z^2+2)(z-3)}$ 16 $\dfrac{z^2 - 3z}{(z+1)^2}$

17 $\dfrac{2z^3 + 3z - 1}{2z^3 + z^2 - 4z + 1}$ 18 $\dfrac{0.205z^3 + 1.388z^2 + 0.875z}{z^3 - 0.603z^2 + 0.185z - 0.436}$

It is suggested that the various methods are used as follows:

Results of Appendix 3	1–10
Partial fractions	11, 13, 14, 15
Division	1–18
Residue	1, 11, 12, 13, 14, 16

10

Solution of difference equations

An important application of the z transform is the solution of linear difference equations with constant coefficients. Difference equations are also referred to as recurrence relations or recurrence equations. These equations arise in the mathematical modeling of general discrete systems and in the simulation of analog systems by digital processing (Shinners, 1964; Kuo, 1975; Borrie, 1986). Such equations are of the form

$$g(n) + a_1 g(n-1) + a_2 g(n-2) + \cdots = b_0 f(n) + b_1 f(n-1) + \cdots$$

where $g(n)$ and $f(n)$ are two sequences. Some typical examples follow.

Digital filter

If i_n is the current in the nth loop and V_n is the voltage at node A_n (Figure 10.1) then the relationship between successive loop currents is

$$i_{n+2} - \left(2 + \frac{Z_1}{Z_2}\right) i_{n+1} + i_n = 0 \tag{10.1}$$

or between successive nodal voltages is

$$V_{n+2} - \left(2 + \frac{Z_1}{Z_2}\right) V_{n+1} + V_n = 0 \tag{10.2}$$

In either case we obtain a difference equation which is linear with constant coefficients. Note that equation (10.2) is a second-order equation since the voltage at node $n + 2$ is expressed as a linear combination of

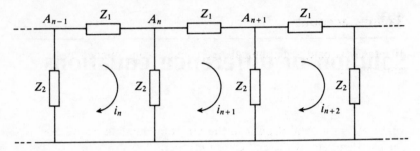

Figure 10.1 A digital filter

the voltages at *two* previous nodes. Similarly the loop equation (10.1) yields a second-order equation in the loop currents.

Simultaneous equations can also be generated. For example

$$y_n = u_n + 5y_{n-1} - 6y_{n-2}$$

$$u_n = x_n + 4u_{n-2}$$

where x_n is the input, y_n is the output and u_n is an intermediate sequence.

Bending moments

The equation satisfied by the bending moments at the supports of a beam resting on several equidistant supports at the same level can be written in the form

$$M_{n+2} + 4M_{n+1} + M_n = \tfrac{1}{2} wa^2$$

where w is the weight per unit length of the beam and a is the distance between the supports.

Population growth

If x_n denotes the size of a population after n years then

$$x_n - x_{n-1} = k_1 x_{n-1} + k_2 x_{n-2}$$

where k_1 and k_2 are constants.

Savings account

If interest is paid at an annual rate r and is compounded N times per annum then

$$y_n = \left(1 + \frac{r}{N}\right) y_{n-1} + x_n$$

where x_n is the total deposit in the nth compounding period and y_n is the total amount at the end of the nth period.

Computing

Difference equations arise in, for example, the modeling of combinatorial problems, selection–sort algorithms and divide and conquer algorithms.

Sampled-data system

If we consider the operation of a sampled-data system where all the signal processing functions can be defined by the operations

1. addition/subtraction,
2. multiplication/division,
3. summation,
4. single-sample delay,

then, since the inputs and outputs are simply strings of numbers, the system can be implemented either as a set of equations or by the use of special-purpose electronic circuits. In either case the system can be defined in terms of elementary building blocks that specify the basic mathematical operations.

From Figure 10.2 the equation is

$$By_n = Cx_n + Ay_{n-1} + Ey_{n-2} + Dx_{n-1}$$

or

$$By_n - Ay_{n-1} - Ey_{n-2} = Cx_n + Dx_{n-1}$$

In order to obtain a solution (y_n in terms of n and x_n) of such an equation we require *two* known values, i.e. initial conditions.

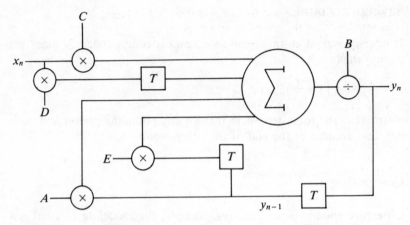

Figure 10.2 A sampled-data system (T = single sample delay)

The method of solution via the z transform is to transform the difference equation using any necessary standard forms (19–22) and obtain an algebraic expression for $F(z)$, the transform of $f(n)$. The function $f(n)$ can be current, voltage or some other function.

Note that the solution $f(n)$ can be derived in a variety of forms depending on the method of inversion that is used but that the result is always a set of sampled values.

EXAMPLE 10.1

Solve the difference equation

$$f(n + 2) - 2f(n + 1) + f(n) = 3n \qquad n > 2$$

given that $f(0) = 0$ and $f(1) = 2$.

Transforming the equation we have

$$[z^2 F(z) - z^2 f(0) - z f(1)] - 2[z F(z) - z f(0)] + F(z) = \frac{3z}{(z - 1)^2}$$

$$F(z)(z^2 - 2z + 1) = \frac{3z}{(z - 1)^2} + 2z$$

$$F(z) = \frac{3z}{(z - 1)^4} + \frac{2z}{(z - 1)^2}$$

Using the residue method for the first term and standard form 4 (Appendix 3) for the second term we obtain

$$f(n) = \tfrac{1}{2}n(n-1)(n-2) + 2n \qquad n > 2$$

EXAMPLE 10.2

Solve the difference equation

$$2f(n) - 3f(n+1) = (-1)^{n-1}$$

if $f(0) = 2$.

Transforming the equation we have

$$2F(z) - 3[zF(z) - zf(0)] = (-1)^{-1}\frac{z}{z+1}$$

$$F(z)(2 - 3z) = -\frac{z}{z+1} - 6z$$

$$F(z) = \frac{z}{(z+1)(3z-2)} + \frac{6z}{3z-2}$$

$$= -\frac{z}{5(z+1)} + \frac{3z}{5(3z-2)} + \frac{6z}{3z-2}$$

in partial fractions. Thus

$$F(z) = -\frac{z}{5(z+1)} + \frac{z}{5(z-2/3)} + \frac{2z}{z-2/3}$$

$$= -\frac{z}{5(z+1)} + \frac{11z}{5(z-2/3)}$$

Hence

$$f(n) = -\tfrac{1}{5}(-1)^n + \tfrac{11}{5}\left(\tfrac{2}{3}\right)^n$$

EXERCISE 10

Solve the following difference equations:

1   $f(n+1) + 2f(n) = (-1)^n$, given that $f(0) = -2$.

2   $2i(n+2) - 3i(n+1) + i(n) = 0$, given that $i(0) = 2$, $i(1) = -1$.

3 $x(n + 2) + 5x(n + 1) + 6x(n) = 3$, given that $x(0) = -2$, $x(1) = 1$.

4 $f(n + 1) - f(n) = n^2$, given that $f(0) = 0$.

5 $f(n + 2) - 4f(n + 1) + 4f(n) = 2^n$, given that $f(0) = 1$, $f(1) = -1$.

6 $x(n + 2) + x(n + 1) - 2x(n) = \sin(n\pi/2)$, given that $x(0) = 0$, $x(1) = 2$.

7 $8y(n + 2) + 6y(n + 1) + y(n) = 5$, given that $y(0) = 0$, $y(1) = -1$.

8 $10f(n + 2) - 11f(n + 1) + 3f(n) = 10$, given that $f(0) = 0$, $f(1) = 0$.

9 $y(n + 2) - \sqrt{3}y(n + 1) + y(n) = 0$, given that $y(0) = 1$, $y(1) = \sqrt{3}$.

10 $f(n + 2) + 9f(n) = 13(2)^{n-2}$, given that $f(0) = 1$, $f(1) = 2$.

11 $i(n + 2) - 4i(n + 1) + i(n) = 0$, given that $i(0) = 0$, $i(1) = -1$.

12 $x(n + 2) - 2x(n + 1) + x(n) = 0$, given that $x(0) = A$, $x(1) = B$.

13 $i(n + 2) - 3i(n + 1) + i(n) = 0$, given that $i(0) = 2$, $i(1) = 6$.

14 $2f(n + 3) - 3f(n + 2) + f(n) = 0$, given that $f(0) = 0$, $f(1) = 1$, $f(2) = -4$.

11
Discrete transfer function

In the same way that differential equations and the Laplace transform enable continuous-time systems to be represented by transfer functions, so can discrete-time systems be represented by transfer functions. We are dealing, of course, with difference equations and the z transform, which then allows discussion of frequency domain concepts such as magnitude and phase response.

Consider a system with input $x(n)$, output $y(n)$ (Figure 11.1) determined by the coefficients a_i, b_j such that the input–output relationship is defined by a difference equation

$$y(n) + b_1 y(n-1) + b_2 y(n-2) + \cdots$$
$$= a_0 x(n) + a_1 x(n-1) + a_2 x(n-2) + \cdots \quad \textbf{(11.1)}$$

Note that the input $x(n)$ is the output of a sampler which is assumed to be ideal. Assuming that all initial conditions are zero, the z transform of equation (11.1) yields

$$Y(z) + b_1 z^{-1} Y(z) + b_2 z^{-2} Y(z) + \cdots$$
$$= a_0 X(z) + a_1 z^{-1} X(z) + a_2 z^{-2} X(z) + \cdots \quad \textbf{(11.2)}$$

where $Y(z) = \mathscr{Z}\{y(n)\}$ and $X(z) = \mathscr{Z}\{x(n)\}$. Collecting terms and rearranging equation (11.2) we obtain

$$\frac{Y(z)}{X(z)} = \frac{a_0 + a_1 z^{-1} + a_2 z^{-2} + \cdots}{1 + b_1 z^{-1} + b_2 z^{-2} + \cdots} = G(z)$$

If both numerator and denominator are of degree k then $G(z)$ can be

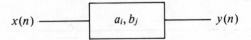

Figure 11.1

written in the form

$$G(z) = \frac{a_0 z^k + a_1 z^{k-1} + a_2 z^{k-2} + \cdots}{z^k + b_1 z^{k-1} + b_2 z^{k-2} + \cdots}$$

with no loss of generality. The numerator can be factorized to obtain the zeros, the denominator to obtain the poles, of $G(z)$ and the system analyzed for stability and response as in Chapter 6.

It should be emphasized, however, that if the output $y(t)$ is a well-behaved function between sampling instants then $y(n)$ and $Y(z)$ will give an accurate description of the true output; otherwise, the z transform may give misleading results.

It should also be noted that there are significant differences between Laplace transforms and z transforms that must be borne in mind. For example, consider two elements in cascade, in one case separated by a sampler and in the other directly linked. Figure 11.2 shows case 1.

$$Y_1(z) = X(z)G_1(z)$$

$$Y_2(z) = Y_1(z)G_2(z)$$

i.e.

$$Y_2(z) = G_1(z)G_2(z)X(z)$$

Hence the overall transfer function is the product of the two separate transfer functions.

Figure 11.3 shows case 2. In this case the second element is driven at the sampling instants and also between sampling instants. The z transform of the output can be shown to be

$$Y_2(z) = X(z)H(z)$$

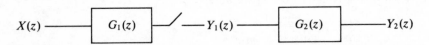

Figure 11.2

Figure 11.3

where $H(z)$ represents the z transform of the cascaded combination corresponding to the transfer function $G_1(s)G_2(s)$.

Note that, in general,

$$G_1(z)G_2(z) \neq H(z)$$

i.e. the product of z transforms is not equal to the z transform of the multiplicands.

For a more detailed discussion of the stability of discrete time systems the reader should consult an appropriate textbook (e.g. Shinners, 1964; Kuo, 1975; Borrie, 1986).

Appendix I
An introduction to the definition of the Laplace transform

This more detailed introduction to the Laplace transform follows that suggested by Hancock.[1] The opportunity is also taken to mention some of the ideas that would be encountered in a more rigorous mathematical development.

Consider the differential equation

$$\frac{d^2 x}{dt^2} - 6 \frac{dx}{dt} + 8x = 9e^{5t}$$

given that $x = 2$, $dx/dt = 9$ at $t = 0$. Suppose that we approach the solution of this equation by considering an integrating factor method.

Let e^{pt} be an integrating factor:

$$e^{pt} \frac{d^2 x}{dt^2} - 6e^{pt} \frac{dx}{dt} + 8xe^{pt} = 9e^{(p+5)t}$$

Now integrate:

$$\int e^{pt} \frac{dx}{dt} \, dt = [e^{pt}x] - p \int e^{pt} x \, dt$$

$$\int e^{pt} \frac{d^2 x}{dt^2} \, dt = e^{pt} \frac{dx}{dt} - p \int e^{pt} \frac{dx}{dt} \, dt$$

$$= e^{pt} \frac{dx}{dt} - p \left(xe^{pt} - p \int xe^{pt} \, dt \right)$$

---

[1] S. T. R. Hancock, 'The Laplace transform', *Mathematical Gazette*, 361 (1963), 215–19.

Substituting into the differential equation we obtain

$$e^{pt}\frac{dx}{dt} - p\left\{xe^{pt} - p\int xe^{pt}\,dt\right\} - 6xe^{pt} + 6p\int xe^{pt}\,dt + 8\int xe^{pt}\,dt$$

$$= 9\int e^{(p+5)t}\,dt$$

$$= \frac{9}{p+5}\,e^{(p+5)t}$$

Collecting terms gives

$$e^{pt}\left(\frac{dx}{dt} - px - 6x\right) + (p^2 + 6p + 8)\int xe^{pt}\,dt = \frac{9}{p+5}\,e^{(p+5)t}$$

This is a complicated equation which appears to be more difficult to solve than the original. Is it therefore possible to simplify this equation?

NOTE 1   The coefficient of $\int xe^{pt}\,dt$ is $p^2 + 6p + 8$. If p were negative, say $-s$, $s > 0$, then $s^2 - 6s + 8$ would be analogous to the left-hand side of the original differential equation

$$\frac{d^2x}{dt^2} - 6\frac{dx}{dt} + 8$$

NOTE 2   If again p were negative the equation can be simplified by choosing to integrate between limits, definite integrals. What limits? Since the initial values are given choose $t = 0$ as the lower limit and, because we have introduced negative exponentials, choose $t = +\infty$ as the upper limit, i.e.

$$\left[e^{-st}\left(\frac{dx}{dt} + sx - 6x\right)\right]_0^\infty + (s^2 - 6s + 8)\int_0^\infty e^{-st}x\,dt$$

$$= \left[\frac{9}{-s+5}\,e^{(-s+5)t}\right]_0^\infty$$

$$0 - (9 + 2s - 12) + (s^2 - 6s + 8)\int_0^\infty e^{-st}x\,dt = 0 - \frac{9}{-s+5}$$

$$(s^2 - 6s + 8)\int_0^\infty e^{-st}x\,dt = \frac{9}{s-5} + 2s - 3$$

$$\int_0^\infty e^{-st}x\,dt = \frac{2s^2 - 13s + 24}{(s-5)(s^2 - 6s + 8)}$$

Note that this is an algebraic expression for $\int_0^\infty e^{-st}x\,dt$ in terms of s. How to obtain x in terms of t is not obvious at this stage but note that the right-hand side can be written in the form

$$\frac{1}{s-2} - \frac{2}{s-4} + \frac{3}{s-5}$$

for which the known solution is

$$x = e^{2t} - 2e^{4t} + 3e^{5t}$$

and that the method operating on the right-hand side of the original differential equation ($9e^{5t}$) produced $9/(s-5)$.

The method, in theory, becomes: multiply by e^{-st}, integrate from 0 to ∞, manipulate algebraically and reverse (invert). We thus define the Laplace transform of a function $f(t)$ as

$$\int_0^\infty e^{-st}f(t)\,dt = \mathscr{L}\{f(t)\} = F(s) \qquad s > 0$$

if this integral exists.

The function e^{-st} is called the *kernel* of the transformation. Other similar transformations are Fourier, Hankel and Mellin.

Existence of the Laplace transform

The definition of the Laplace transform is an improper integral which may or may not converge. It is possible then that the Laplace transform of certain functions may not exist or that the inverse may not be unique. The sufficient conditions for the existence of the transform are stated in theorem 1.

THEOREM 1 If $f(t)$ is piecewise continuous for $t \geqslant 0$ and satisfies the condition

$$|f(t)| \leqslant Me^{at}, t \geqslant T$$

where a, M and T are fixed non-negative constants then $\mathscr{L}\{f(t)\}$ exists for all $s > a$.

The theorem implies that all bounded continuous functions and every polynomial have Laplace transforms. Thus t^n, $n = 1, 2, 3, \ldots$ has a Laplace transform but e^{t^2} does not. It should also be noted that the conditions are not necessary for the existence. There are functions, such as $t^{-1/2}$, that do not satisfy the conditions but yet have a Laplace transform.

$$\mathscr{L}\{t^{-1/2}\} = \int_0^\infty e^{-st}t^{-1/2}\,dt = \frac{\Gamma(\tfrac{1}{2})}{s^{1/2}} = \sqrt{\frac{\pi}{s}}$$

It can be shown that functions that differ only at several isolated points can in fact have the same Laplace transform. This loss of uniqueness can only be accommodated by the inclusion of null functions. These functions do not generally arise in practical situations so that their exclusion presents no difficulty in the application of Laplace transforms. The uniqueness is then stated in theorem 2.

THEOREM 2   Lerch's Theorem.   If $f(t)$ is piecewise continuous for $t \geqslant 0$ and satisfies the condition

$$|f(t)| \leqslant Me^{at} \qquad t \geqslant T$$

where a, M and T are fixed non-negative constants then the inverse Laplace transform of $F(s)$ where $f(t) = \mathscr{L}^{-1}\{F(s)\}$, is unique.

Hence we assume that the Laplace transform and the inverse, for all functions of practical interest, are unique.

Appendix 2
A table of Laplace transforms

	$f(t) = \mathscr{L}^{-1}\{F(s)\}$	$F(s) = \mathscr{L}\{f(t)\} = \displaystyle\int_0^\infty e^{-st} f(t)\, dt$
1 Sum	$af(t) + bg(t)$	$aF(s) + bG(s)$ or $a\bar{f}(s) + b\bar{g}(s)$
2 First derivative	$\dfrac{d}{dt} f(t)$ or $f'(t)$	$sF(s) - f(0)$ or $s\bar{f}(s) - f_0$
3 Second derivative	$\dfrac{d^2}{dt^2} f(t)$ or $f''(t)$	$s^2 F(s) - sf(0) - f'(0)$ or $s^2\bar{f}(s) - sf_0 - f_1$
4 Third derivative	$\dfrac{d^3}{dt^3} f(t)$	$s^3 F(s) - s^2 f(0) - sf'(0) - f''(0)$
5 Fourth derivative	$\dfrac{d^4}{dt^4} f(t)$	$s^4 F(s) - s^3 f(0) - s^2 f'(0) - sf''(0) - f'''(0)$

(Continued)

103

	$f(t) = \mathcal{L}^{-1}\{F(s)\}$	$F(s) = \mathcal{L}\{f(t)\} = \int_0^\infty e^{-st} f(t)\, dt$
6 Definite integral	$\displaystyle\int_0^t f(t)\, dt$	$\dfrac{1}{s} F(s)$ or $\dfrac{1}{s}\bar{f}(s)$
	$\displaystyle\int_{-\infty}^t f(t)\, dt$	$\dfrac{1}{s} F(s) + \dfrac{1}{s}\displaystyle\int_{-\infty}^0 f(t)\, dt$
7 Exponential multiplier	$e^{-\alpha t} f(t)$	$F(s+\alpha)$ or $\bar{f}(s+\alpha)$
8 Time shift	$f(t-T)u(t-T)$	$e^{-sT} F(s)$ or $e^{-sT}\bar{f}(s)$
9 Periodic function	$f(t) = f(t+T)$	$\dfrac{1}{1-e^{-sT}}\displaystyle\int_0^T e^{-st} f(t)\, dt$
10 Convolution	$f(t) * g(t) = \displaystyle\int_0^t f(t-u)g(u)\, du$	$F(s)G(s)$ or $\bar{f}(s)\bar{g}(s)$
11 Unit step	$u(t)$ or $H(t)$	$\dfrac{1}{s}$
12 Delayed step	$u(t-T)$	$\dfrac{1}{s}e^{-sT}$

			$f(t)$	$F(s)$
13	Unit impulse		$\delta(t)$	1
14	Delayed unit impulse		$\delta(t-a)$	e^{-as}
15	Linear ramp		t	$\dfrac{1}{s^2}$
16	nth order ramp		t^n	$\dfrac{n!}{s^{n+1}}$
17	Exponential decay		$e^{-\alpha t}$	$\dfrac{1}{s+\alpha}$
18		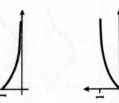	$\delta(t)-\alpha e^{-\alpha t}$	$\dfrac{s}{s+\alpha}$
19			$1-e^{-\alpha t}$	$\dfrac{\alpha}{s(s+\alpha)}$

(Continued)

	$f(t) = \mathcal{L}^{-1}\{F(s)\}$	$F(s) = \mathcal{L}\{f(t)\} = \int_0^\infty e^{-st} f(t)\, dt$
20	$te^{-\alpha t}$	$\dfrac{1}{(s+\alpha)^2}$
21	$t^n e^{-\alpha t}$	$\dfrac{n!}{(s+\alpha)^{n+1}}$
22	$(1-\alpha t)e^{-\alpha t}$	$\dfrac{s}{(s+\alpha)^2}$
23	$e^{-\alpha t} - e^{-\beta t}$	$\dfrac{\beta - \alpha}{(s+\alpha)(s+\beta)}$
24	$\dfrac{\alpha e^{-\alpha t} - \beta e^{-\beta t}}{\alpha - \beta}$	$\dfrac{s}{(s+\alpha)(s+\beta)} \qquad \alpha \neq \beta$
25 Sine wave	$\sin(\omega t)$	$\dfrac{\omega}{s^2 + \omega^2}$
26	$\sin(\omega t \pm \phi)$	$\dfrac{\omega \cos(\phi) \pm s \sin(\phi)}{s^2 + \omega^2}$

(Continued)

| 27 | | | $t \sin(\omega t)$ | $\dfrac{2\omega s}{(s^2 + \omega^2)^2}$ |

| 28 | Cosine wave | | $\cos(\omega t)$ | $\dfrac{s}{s^2 + \omega^2}$ |

| 29 | |  | $\cos(\omega t \pm \phi)$ | $\dfrac{s\cos(\phi) \mp \omega\sin(\phi)}{s^2 + \omega^2}$ |

| 30 | | | $t \cos(\omega t)$ | $\dfrac{s^2 - \omega^2}{(s^2 + \omega^2)^2}$ |

| 31 | |  | $1 - \cos(\omega t)$ | $\dfrac{\omega^2}{s(s^2 + \omega^2)}$ |

			$f(t) = \mathscr{L}^{-1}\{F(s)\}$	$F(s) = \mathscr{L}\{f(t)\} = \int_0^\infty e^{-st} f(t)\, dt$
32			$\sin(\omega t) - t\cos(\omega t)$	$\dfrac{2\omega^3}{(s^2+\omega^2)^2}$
33	Exponentially damped		$e^{-\alpha t}\sin(\omega t)$	$\dfrac{\omega}{(s+\alpha)^2+\omega^2}$
34			$e^{-\alpha t}\cos(\omega t)$	$\dfrac{s+\alpha}{(s+\alpha)^2+\omega^2}$
35			$e^{-\alpha t}[\sin(\omega t) - \omega t\cos(\omega t)]$	$\dfrac{2\omega^3}{[(s+\alpha)^2+\omega^2]^2}$
36	Hyperbolic function		$\sinh(\omega t)$	$\dfrac{\omega}{s^2-\omega^2}$
37			$\cosh(\omega t)$	$\dfrac{s}{s^2-\omega^2}$
38	Damped hyperbolic		$e^{-\alpha t}\sinh(\omega t)$	$\dfrac{\omega}{(s+\alpha)^2-\omega^2}$

| 39 | $e^{-\alpha t}\cosh(\omega t)$ | $\dfrac{s+\alpha}{(s+\alpha)^2 - \omega^2}$ |

| 40 | $e^{-\alpha t}[\sinh(\omega t) - \omega t\,\cosh(\omega t)]$ | $\dfrac{-2\omega^3}{[(s+\alpha)^2 - \omega^2]^2}$ |

41 (a) $\zeta < 1$ and $\omega_d = \omega_0(1 - \zeta^2)^{1/2}$ where ω_d is the frequency of free damped oscillation.

$$u(t) - e^{-\zeta\omega_0 t} \times \left[\cos(\omega_d t) + \dfrac{\zeta\omega_0}{\omega_d}\sin(\omega_d t)\right]$$

$$\left[s\left(\dfrac{s^2}{\omega_0^2} + \dfrac{2\zeta s}{\omega_0} + 1\right)\right]^{-1}$$

where ω_0 is the frequency of undamped oscillations, i.e. if $\zeta = 0$

(b) $\zeta = 1$ $u(t) - e^{-\omega_0 t}[1 + \omega_0 t]$

(c) $\zeta > 1$ and $\beta = \omega_0(\zeta^2 - 1)^{1/2}$

$$u(t) - e^{-\zeta\omega_0 t} \times \left[\cosh(\beta t) + \dfrac{\zeta\omega_0}{\beta}\sinh(\beta t)\right]$$

42 Delayed ramp

$(t - T)u(t - T)$ $\dfrac{1}{s^2}e^{-sT}$

43 Rectangular pulse

$u(t) - u(t - T)$ $\dfrac{1}{s}(1 - e^{-sT})$

44 Rectangular periodic wave, period T

$f(t) = \begin{cases} 1 & 0 < t < T/2 \\ 0 & T/2 < t < T \end{cases}$ $\dfrac{1}{s(1 + e^{-sT/2})}$

(Continued)

$f(t) = \mathscr{L}^{-1}\{F(s)\}$	$F(s) = \mathscr{L}\{f(t)\} = \int_0^\infty e^{-st} f(t)\, dt$

45 Half-wave-rectified sine, period $T = 2\pi/\omega$

$$f(t) = \begin{cases} \sin(\omega t) & 0 < t < T/2 \\ 0 & T/2 < t < T \end{cases}$$

$$\frac{\omega}{(s^2 + \omega^2)(1 - e^{-\pi s/\omega})}$$

46 Full-wave-rectified sine, period $T = 2\pi/\omega$

$$f(t) = |\sin(\omega t)|$$

$$\frac{\omega}{s^2 + \omega^2} \frac{(1 + e^{-\pi s/\omega})}{(1 - e^{-\pi s/\omega})}$$

Initial value theorem $\displaystyle \lim_{t \to 0} f(t) = \lim_{s \to \infty} sF(s)$ Final value theorem $\displaystyle \lim_{t \to \infty} f(t) = \lim_{s \to 0} sF(s)$

Appendix 3
A table of z transforms

$f(n)$		$F(z) = \sum\limits_{n=0}^{\infty} f(n)z^{-n}$
1 $f(0) = k$ $f(n) = 0, \; n = 1, 2, \ldots$	Impulse at $n = 0$	k
2 $f(m) = k$ $f(n) \; = 0, \; n \neq m$	Impulse at $n = m$	kz^{-m}
3 k		$\dfrac{kz}{z-1}$
4 kn		$\dfrac{kz}{(z-1)^2}$
5 kn^2		$\dfrac{kz(z+1)}{(z-1)^3}$
6 ke^{-an}		$\dfrac{kz}{z-e^{-a}}$
7 kne^{-an}		$\dfrac{kze^{-a}}{(z-e^{-a})^2}$
8 $\sin(\omega_0 n)$		$\dfrac{z\sin(\omega_0)}{z^2 - 2z\cos(\omega_0) + 1}$

(*Continued*)

$f(n)$	$F(z) = \sum\limits_{n=0}^{\infty} f(n)z^{-n}$
9 $\cos(\omega_0 n)$	$\dfrac{z[z - \cos(\omega_0)]}{z^2 - 2z\cos(\omega_0) + 1}$
10 $e^{-an}\sin(\omega_0 n)$	$\dfrac{ze^{-a}\sin(\omega_0)}{z^2 - 2ze^{-a}\cos(\omega_0) + e^{-2a}}$
11 $e^{-an}\cos(\omega_0 n)$	$\dfrac{ze^{-a}[ze^{a} - \cos(\omega_0)]}{z^2 - 2ze^{-a}\cos(\omega_0) + e^{-2a}}$
12 α^n where α is constant	$\dfrac{z}{z - \alpha}$
13 $n\alpha^n$	$\dfrac{\alpha z}{(z - \alpha)^2}$
14 $\sinh(\omega_0 n)$	$\dfrac{z\sinh(\omega_0)}{z^2 - 2z\cosh(\omega_0) + 1}$
15 $\cosh(\omega_0 n)$	$\dfrac{z[z - \cosh(\omega_0)]}{z^2 - 2z\cosh(\omega_0) + 1}$
16 $f(0) = 0$ $f(n) = 1/n \qquad n \geqslant 1$	$\ln\left(\dfrac{1}{z - 1}\right)$
17 $f(0) = 0$ $f(n) = \dfrac{(-1)^{n-1}}{n} \qquad n \geqslant 1$	$\ln\left(\dfrac{z + 1}{z}\right)$
18 $\dfrac{\alpha^n}{n!}$	$e^{\alpha/z}$
19 $f(n + 1)$	$zF(z) - zf(0)$
20 $f(n + 2)$	$z^2 F(z) - z^2 f(0) - zf(1)$
21 $f(n + 3)$	$z^3 F(z) - z^3 f(0) - z^2 f(1) - zf(2)$
22 $f(n + m) \qquad m \geqslant 0$	$z^m F(z) - z^m f(0) - \cdots - zf(m - 1)$

(Continued)

$f(n)$		$F(z) = \sum\limits_{n=0}^{\infty} f(n)z^{-n}$
23 $f(n-m)u(n-m)$	$m \geqslant 0$	$z^{-m}F(z)$
24 $nf(n)$		$-z\,\dfrac{\mathrm{d}[F(z)]}{\mathrm{d}z}$
25 $\dfrac{1}{n}f(n)$		$-\displaystyle\int \dfrac{F(z)}{z}\,\mathrm{d}z$
26 $\dfrac{1}{n+m}f(n)$		$-z^{m}\displaystyle\int \dfrac{F(z)}{z^{m+1}}\,\mathrm{d}z$
27 $a^{n}f(n)$		$F(z/a)$
28 $\sum\limits_{k=0}^{n} f(k)$		$\dfrac{zF(z)}{z-1}$
29 $\sum\limits_{k=0}^{n} f_2(k)f_1(n-k)$		$F_1(z)F_2(z)$

Initial value theorem $\underset{n\to 0}{\text{limit}}\; f(n) = \underset{z\to\infty}{\text{limit}}\; F(z)$

Final value theorem $\underset{n\to\infty}{\text{limit}}\; f(n) = \underset{z\to 1}{\text{limit}}\; \dfrac{z-1}{z} F(z)$

Appendix 4
Types of sampling

Instantaneous sampling is not feasible. It is not possible to construct switches which can operate in an arbitrarily short time. It is common practice therefore to use either natural or flat-topped sampling.

Natural sampling

The sampling waveform consists of a train of pulses of duration τ and separated by the sampling time T. The sampled signal then consists of a sequence of pulses of varying magnitude whose tops follow the waveform of $m(t)$. (Figure A4.1.)

A low-pass filter will deliver an output signal which is directly proportional to τ and inversely proportional to T so that there must be a compromise between large τ and a large guard time between the end of one sample and the beginning of another.

Flat-topped sampling

More frequently flat-topped pulses are used. A flat-topped pulse has a constant amplitude which is established by the sample value of the signal at some point within the pulse interval.

Figure A4.2 shows sampling of the signal at the beginning of the pulse. In sampling of this type the signal $m(t)$ cannot be recovered exactly but the distortion need not be large. Flat-topped sampling has

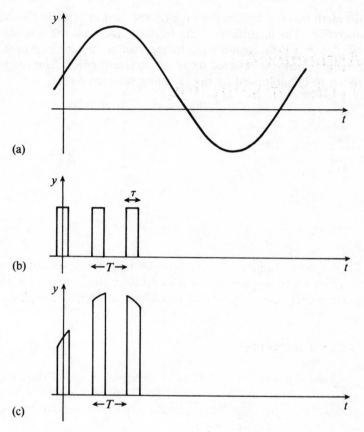

Figure A4.1 (a) A signal $y = m(t)$; (b) sampling waveform; (c) naturally sampled signal

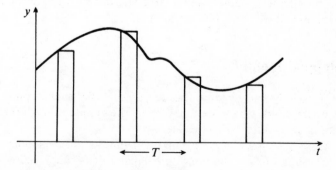

Figure A4.2 Flat-topped sampling

the merit that it simplifies the circuitry required to perform the sampling operation. The distortion results from the fact that the original signal was 'observed' through a finite rather than an infinitesimal time 'aperture' and hence is referred to as the *aperture effect*. The spectrum is found to be multiplied by the sampling function $Sa(x)$.

Answers to exercises

Exercise 1.1

1 $\dfrac{5}{s^2} - \dfrac{2}{s}$

2 $\dfrac{a}{s} + \dfrac{b}{s^2} + \dfrac{2c}{s^3}$

3 $\dfrac{6}{s^4} + \dfrac{8}{s+1} + \dfrac{1}{s}$

4 $\dfrac{4}{s^3} - \dfrac{6}{s^2}$

5 $\dfrac{2}{s} + \dfrac{4}{s-3}$

6 $\dfrac{1}{s+a} + \dfrac{1}{s+b}$

7 $\dfrac{a^2}{s^2+a^2} + \dfrac{b^2}{s^2+b^2}$

8 $\dfrac{s}{s^2+a^2} + \dfrac{s}{s^2+b^2}$

9 $\dfrac{10}{s^3} + \dfrac{4s}{s^2+9}$

10 $\dfrac{3}{s^2-9}$

11 $\dfrac{s}{s^2-9}$

12 $\dfrac{s^2-4}{(s^2+4)^2} - \dfrac{s}{s^2-16}$

13 $\dfrac{s+3}{s^2+6s+13}$

14 $\dfrac{24}{(s-3)^5} - \dfrac{1}{s^2+4s+5}$

15 $\dfrac{8}{s(s^2-4)}, \dfrac{2}{s(s^2-4)}$

16 $\dfrac{s-2}{s^2-4s+3}$

17 $\dfrac{s\cos(\alpha) + a\sin(\alpha)}{s^2+a^2}$

18 $\dfrac{b\cos(\alpha) + s\sin(\alpha)}{s^2+b^2}$

117

19 $\dfrac{2}{s^2 + 16} - \dfrac{1}{s^2 + 4}$ **20** $\dfrac{s}{2(s^2 + 9)} + \dfrac{s}{2(s^2 + 1)}$

21 $\dfrac{1}{2s} - \dfrac{s}{2(s^2 + 4)}$

Exercise 1.2

1 $\frac{1}{6}t^3$ **2** $t + \frac{1}{2}t^2$

3 $2t - 3 + 5e^{-t}$ **4** $\frac{1}{2}e^{-t/2}$

5 $\frac{1}{3}e^{-4t/3}$ **6** $\frac{1}{4}\cos\left(\dfrac{t}{2}\right)$

7 $\frac{1}{2}\sinh(2t)$ **8** $9\cosh(4t)$

9 $2\cos(3t) - \frac{5}{3}\sin(3t)$ **10** $\cos(t) + 3\sin(t)$

11 $\frac{3}{5}e^{2t} - \frac{3}{5}e^{-3t}$ **12** $\frac{1}{3}e^{-3t}\sin(3t)$

13 $\frac{1}{6}t\sin(3t)$ **14** $\frac{5}{128}[\sin(4t) - 4t\cos(4t)]$

15 $t^7 e^{3t}$ **16** $\frac{3}{4}e^{-3t} + \frac{1}{4}e^{t}$

17 $e^{-t}\cosh(t\sqrt{3}) - \dfrac{1}{\sqrt{3}}e^{-t}\sinh(t\sqrt{3})$

18 $e^{-3t}[\frac{1}{16}\sin(2t) - \frac{1}{8}t\cos(2t)]$

19 $\frac{2}{13}e^{2t} - \frac{2}{13}\cos(3t) + \frac{35}{39}\sin(3t)$

20 $\frac{23}{26}e^{3t} - \frac{19}{26}e^{-3t} - \frac{2}{13}\cos(2t) + \frac{14}{13}\sin(2t)$

21 $\frac{1}{4}t - \frac{1}{3}\sin(t) + \frac{1}{24}\sin(2t)$

Exercise 2

1 $y = \frac{1}{4} + \frac{7}{4}e^{-4t}$

2 $y = \frac{1}{2} + \frac{9}{2}e^{-2t} - 3e^{-3t}$

3 $y = \frac{9}{2}te^{-t} + \frac{7}{2}e^{-t} - \frac{1}{2}\cos(t)$

4 $T = T_0 e^{\mu\theta}$

5 $r = (1 + 4\theta)e^{-2\theta}$

6 $x = \cos(4t) + \frac{1}{2}\sin(4t) + 2$

7 $x = \frac{5}{7}e^{-4t} - \frac{3}{4}e^{-t} + \frac{1}{28}e^{3t}$

8 $x = \frac{14}{25}e^{-t} + \frac{1}{5}te^{-t} - \frac{39}{25}e^{-6t}$, $y = \frac{24}{25}e^{-t} + \frac{26}{25}e^{-6t} + \frac{1}{5}te^{-t}$

9 $x = 3e^t - \frac{9}{5}e^{-t} - \frac{6}{5}\cos(2t) - \frac{12}{5}\sin(2t)$,

   $y = 3e^{-t} - 3e^t + 3\sin(2t)$

10 $x = \frac{3}{2}e^t + \frac{11}{10}e^{-3t} + \frac{2}{5}e^{2t} - 1$

   $y = \frac{3}{2}e^t - \frac{11}{10}e^{-3t} + \frac{3}{5}e^{2t} - 2$

11 $x = (A + Bt)e^t + (E + Ft)e^{-t}$

   $y = \frac{1}{2}(B - A - Bt)e^t - \frac{1}{2}(E + F + Ft)e^{-t}$

12 $i_1 = 8\sin(t) + 6\cos(t) - e^{-t/3} - 5e^{-t}$

   $i_2 = 2\sin(t) + 4\cos(t) + e^{-t/3} - 5e^{-t}$

Exercise 3

1 $\dfrac{2\omega s}{(s^2 + \omega^2)^2}$

2 $\dfrac{2s^3 - 6\omega^2 s}{(s^2 + \omega^2)^3}$

3 $\dfrac{8 + 12s - 8s^2}{(s^2 + 1)^2}$

4 $\dfrac{s^2 + 4}{(s^2 - 4)^2}$

5 $\dfrac{8s}{(s^2 - 16)^2}$

6 $\dfrac{3(2s + 4)}{s(s^2 + 4s + 13)^2}$

7 $\dfrac{1}{s}\cot^{-1}\left(\dfrac{s + 2}{3}\right)$

8 $\ln\left(\dfrac{s + a}{s + b}\right)$

9 $\dfrac{1}{2}\ln\left(\dfrac{s + 1}{s - 1}\right)$

10 $\dfrac{1}{(1 - e^{-\pi s})(s^2 + 1)}$

11 $\dfrac{1 - e^{-s}(s + 1)}{s^2(1 - e^{-2s})}$

12 $\dfrac{k}{s^2} - \dfrac{kTe^{-sT}}{s(1 - e^{-sT})}$

13 $\dfrac{1}{t}(1 - e^{-t})$ **14** $\dfrac{1}{t}(1 - e^{-t})$

15 $\displaystyle\int_0^t \dfrac{1}{t}(1 - e^{-t})\,dt$ **16** $\dfrac{1}{t}(e^{-bt} - e^{-at})$

17 $\displaystyle\int_0^t \dfrac{1}{t}(e^{-bt} - e^{-at})\,dt$ **18** $e^{-t} + t - 1$

19 $1 - e^{-t}(1 + t)$

Exercise 4.1

1 $\dfrac{1}{s}(4 - e^{-s})$ **2** $\dfrac{1}{s}(4 - 2e^{-2s})$

3 $\dfrac{1}{s}(e^{-s} + 3e^{-4s})$ **4** $\dfrac{1}{s}(2 - e^{-3s})$

5 $\dfrac{e^{-s}}{s}(1 - 2e^{-s} + 4e^{-2s})$ **6** $\dfrac{5}{s}e^{-3s}$

7 $\dfrac{1}{s}(3 - 4e^{-s})$ **8** $\dfrac{2e^{-s}}{s} - \dfrac{4e^{-3s}}{s}$

9 $y = 2u(t - 1) - u(t - 3)$, $\bar{y} = \dfrac{1}{s}(2e^{-s} - e^{-3s})$

10 $y = u(t) + 2u(t - 2) - u(t - 3)$, $\bar{y} = \dfrac{1}{s}(1 + 2e^{-2s} - e^{-3s})$

11 $y = u(t - 1) + u(t - 2) - u(t - 3) - u(t - 4)$, $\bar{y} = \dfrac{1}{s}(e^{-s} + e^{-2s} - e^{-3s} - e^{-4s})$

12 $\dfrac{1 - e^{-2(s+3)}}{s + 3}$

13 (a) $2(t - 1)u(t - 1)$

(b) $\cos(t) + [\cos(2t) - \cos(t)]u(t - \pi) + [\cos(3t) - \cos(2t)]u(t - 2\pi)$

14 $\dfrac{2e^{-3s}}{s^2}(1 + 3s)$ **15** $\dfrac{e^{-(s-3)}}{s - 3}$

16 $\dfrac{se^{-\pi s/3}}{s^2 + 1}$ **17** $\dfrac{6e^{-s}}{s^4}$

18 $\dfrac{e^{-(s-3)}}{s-3}$

20 $\bar{y} = \dfrac{1}{s(1-e^{-sT})}$

Exercise 4.2

1 $\cos[3(t-1)]u(t-1)$

2 $\{\frac{1}{2}e^{2(t-1)} + \frac{1}{2}\cos[2(t-1)] + \frac{1}{2}\sin[2(t-1)]\}u(t-1)$

3 $\frac{1}{6}(t-2)^3 e^{-(t-2)}u(t-2) - \cos[2(t-1)]u(t-1)$

4 $\frac{1}{3}\sin[3(t-3)]u(t-3)$

5 $\frac{1}{2}(t-1)^2 e^{-2(t-1)}u(t-1)$

6 $4\sin[2(t-2)]u(t-2)$

7 $\sinh(t-\pi/6)u(t-\pi/6)$

8 $\frac{1}{6}(t-5)^3 e^{3(t-5)}u(t-5)$

9 $[1-\cos(t-2)]u(t-2) - [1-\cos(t-3)]u(t-3)$

10 $i = [5e^{-(t-1)} + e^{5(t-1)} - 6]u(t-1) + e^{5t} - e^{-t}$

11 $x = \frac{1}{2}(1 - 2e^{-(t-2)} + e^{-2(t-2)})u(t-2)$

12 $x = 2 - \cos(t) - 2[1-\cos(t-3)]u(t-3)$

13 $y = [8e^{-(t-1)} - 8te^{-2(t-1)}]u(t-1) - (1+t)e^{-2t}$

14 $i = \frac{25}{16}[1-\cos(4t)] - \frac{25}{16}\{1-\cos[4(t-2)]\}u(t-2)$

15 $x = \frac{3}{2}\sin(2t) + \frac{1}{4}t\sin(2t) - \frac{1}{4}(t-\pi)\sin[2(t-\pi)]u(t-\pi)$

16 $i_2 = \dfrac{100}{\sqrt{6}}\left\{e^{-2t}\sinh\left(\sqrt{\dfrac{2}{3}}\,t\right) - e^{-2(t-1)}\sinh\left[\sqrt{\dfrac{2}{3}}\,(t-1)\right]u(t-1)\right\}$

Exercise 5

1 A

2 $8e^{3cs}$

3 $F_0 e^{-3s}$

4 $\frac{5}{2}e^{-5s/2}$

5 $x = \cos(4t) + \dfrac{P_0}{4} \sin(4t)$ **6** $Y = 2 \sin(3t)$

7 $Y = \cos(2t) + \frac{1}{2} \sin[2(t-1)]\, u(t-1)$

8 $q = E_0 C(1 - e^{-t/RC}) + \dfrac{E_1}{R} e^{-(t-1)/RC} u(t-1)$

9 $x = \dfrac{P_0}{m\omega} e^{-\alpha(t-t_0)} \sin[\omega(t-t_0)]\, u(t-t_0)$

$\quad\quad\quad$ where $\alpha = c/2m$ and $\omega^2 = k - c^2/4m^2$

10 $EIy = \dfrac{W}{6}(x - l/3)^3 u(x - l/3) - \dfrac{W}{9} x^3 + \dfrac{5}{81} W l^2 x$

11 $i = \dfrac{E_0}{2}[(e^{-t} - e^{-3t}) + (e^{-(t-1)} - e^{-3(t-1)})u(t-1)$

$\quad\quad\quad\quad\quad\quad + (e^{-(t-2)} - e^{-3(t-2)})u(t-2)\ldots]$

Exercise 6

	Poles	Zeros	Magnitude	Phase
1	$s = 0$	$s = -1$	$\dfrac{\sqrt{(1 + \omega^2)}}{\omega^3}$	$\dfrac{\pi}{2} + \tan^{-1}(\omega)$
2	$s = \pm 4$	$s = 0$	$\dfrac{9\omega}{\omega^2 + 16}$	$\dfrac{3\pi}{2}$
3	$s = -1/2$	–	$\dfrac{1}{\sqrt{(1 + 4\omega^2)}}$	$-\tan^{-1}(2\omega)$
4	$s = \pm j$	$s = -3$	$\dfrac{\sqrt{(9 + \omega^2)}}{1 - \omega^2}$	$\tan^{-1}(\omega/3)$
5	$s = \pm j\,\frac{1}{2}$	$s = 0$	$\dfrac{\omega}{1 - 4\omega^2}$	$\pi/2$
6	$s = -1 \pm j$	$s = 0$	$\dfrac{\omega}{\sqrt{(4 + \omega^4)}}$	$\dfrac{\pi}{2} - \tan^{-1}\left(\dfrac{2\pi}{2 - \omega^2}\right)$

Exercise 7

1 (a) $5t^2$ $\quad t \geqslant 0$ $\hspace{3cm}$ (b) $\frac{1}{4}t^4 - t^3$ $\quad t \geqslant 0$

(c) $-\frac{3}{8}\cos(3t) + \frac{3}{8}\cos(t)$ $\quad t \geqslant 0$ $\hspace{1.5cm}$ (d) $1 + t - e^{-t}$ $\quad t \geqslant 0$

(e) $t - 1 + e^{-t} - (t-1)u(t-1)$

(f) $2 - 2\cos(2t) - \{2 - 2\cos[2(t-2)]\}u(t-2)$

2 (a) $3e^{-3t} * e^{-t}$

(b) $5\cos(3t) * e^{4t}$

(c) $u(t-3) * \frac{1}{4}\sin(4t)$

(d) $\delta(t-2) * e^{-3t}$

(e) $[2u(t-1) - 3u(t-2)] * \cos(3t)$

3 (a) $\cos(4t)$ $\hspace{4cm}$ (b) $e^{-2(t-2)}u(t-2)$

4 (a) $y = \displaystyle\int_0^t (t-u)e^{-a(t-u)}f(u)\,\mathrm{d}u$ $\quad$ or $\quad$ $y = \displaystyle\int_0^t ue^{-au}f(t-u)\,\mathrm{d}u$

(b) $i = f(t) * \dfrac{1}{b-a}(e^{-at} - e^{-bt}) = \displaystyle\int_0^t f(t-u)\dfrac{e^{-au} - e^{-bu}}{b-a}\,\mathrm{d}u$

(c) $x = 2e^{-2t} - 2e^{-3t} + \displaystyle\int_0^t \dfrac{1}{1+u}(e^{-2(t-u)} - e^{-3(t-u)})\,\mathrm{d}u$

5 $\theta = \displaystyle\int_0^t K\phi(u)\sin[K(t-u)]\,\mathrm{d}u$ $\quad$ or $\quad$ $\displaystyle\int_0^t K\phi(t-u)\sin(Ku)\,\mathrm{d}u$

6 (a) $x = -t + e^t - e^{-t}$

(b) $x = 3(e^t + e^{-t}) * \sin(t) = \frac{3}{2}e^t + \frac{3}{2}e^{-t} - 3\cos(t)$

Exercise 8

1 $\dfrac{z}{(z-1)^2}$ $\hspace{2cm}$ **2** $\dfrac{z(z+1)}{(z-1)^3}$ $\hspace{2cm}$ **3** $\dfrac{3z}{3z+2}$

4 $\dfrac{6z}{(z-1)(z+5)}$ $\hspace{1.3cm}$ **5** $\dfrac{ze^{-4}}{(z-e^{-4})^2}$ $\hspace{1.3cm}$ **6** $\dfrac{2z}{z-1} + \dfrac{3z}{z-e^{-2}}$

7 $\dfrac{z^2}{z^2+1}$ $\hspace{2.2cm}$ **8** $\dfrac{ze^{-a}\sin(\omega)}{z^2 - 2ze^{-a}\cos(\omega) + e^{-2a}}$

9 $\dfrac{3z\,\sin(2)}{z^2 - 2z\,\cos(2) + 1} - \dfrac{4z\,[z - \cos(3)]}{z^2 - 2z\,\cos(3) + 1}$

10 $\dfrac{2.728z}{z^2 + 2.497z + 9}$ **11** $\dfrac{z}{z - 1} - \dfrac{4}{z - 1}$ **12** $\dfrac{z}{z - e^{-T}} + \dfrac{z}{z + e^{-T}}$

13 $\dfrac{z^{1-k}}{z - 1}$ **14** $\dfrac{z\,[z - \cosh(a)]}{z^2 - 2z\,\cosh(a) + 1}$

15 $\dfrac{2z}{(z - 1)^2} - \dfrac{z}{z - 1} + \dfrac{3z}{z - e^{-1}}$

16 $\dfrac{2z - 3}{(z - 1)^2}$ **17** $\dfrac{1 - e^{-1}}{z - e^{-1}}$

Exercise 9

1 $\dfrac{1}{3} n(3)^n$ **2** $\sin\left(\dfrac{n\pi}{2}\right)$ **3** $\cos\left(\dfrac{n\pi}{2}\right)$

4 $\dfrac{2}{\sqrt{3}} \sin\left(\dfrac{n\pi}{3}\right)$ $0,\ 1,\ 1,\ 0,\ -1,\ldots$

5 $\cos\left(\dfrac{n\pi}{3}\right) + \dfrac{1}{\sqrt{3}} \sin\left(\dfrac{n\pi}{3}\right)$ $1,\ 1,\ 0,\ -1,\ -1,\ldots$

6 $\dfrac{1}{\sqrt{3}} \sinh(\omega_0 n)$ where $\omega_0 = \cosh^{-1}(2)$ $0,\ 1,\ 4,\ 15,\ldots$

7 $\cosh(\omega_0 n) + \dfrac{2}{\sqrt{3}} \sinh(\omega_0 n)$ where $\omega_0 = \cosh^{-1}(2)$ $1,\ 4,\ 15,\ldots$

8 $\left(\dfrac{1}{2}\right)^{(n-2)} \sin\left(\dfrac{n\pi}{6}\right)$

9 $\left(\dfrac{1}{2}\right)^n \cos\left(\dfrac{n\pi}{6}\right) + \sqrt{3}\left(\dfrac{1}{2}\right)^n \sin\left(\dfrac{n\pi}{6}\right)$

10 $-(n - 3)(-2)^{(n-3)}u(n - 3)$ $0,\ 0,\ 0,\ 0,\ 2,\ -8,\ 24,\ldots$

11 $-\dfrac{1}{3}(-1)^n - n(-1)^n + \dfrac{1}{3}(2)^n$

12 $(n^2 + 3n + 2)(2)^n$ **13** $\dfrac{1}{7}(-3)^n + \dfrac{6}{7}(4)^n$

14 $-\dfrac{2}{3}(-2)^n + \dfrac{2}{3}$ $0,\ 2,\ -2,\ 6,\ -10,\ldots$

15 $(3)^n - (2)^{n/2} \cos\left(\dfrac{n\pi}{2}\right)$ $0, 3, 11, 27, \ldots$

16 $(4n + 1)(-1)^n$ $1, -5, 9, \ldots$

17 $1, -0.5, 3.75, -3.875, \ldots$ **18** $0.205, 1.512, 1.749, 0.864, \ldots$

Note that the sequence values may be obtained from the function $f(n)$ by substitution of successive values of $n = 0, 1, 2, 3, \ldots$.

Exercise 10

1 $f(n) = (-1)^n - 3(-2)^n$

2 $i(n) = 6(2)^{-n} - 4$

3 $x(n) = \frac{1}{4} - 6(-2)^n + \frac{15}{4}(-3)^n$

4 $f(n) = \frac{1}{6}n(n - 1)(2n - 1)$

5 $f(n) = (2)^{n-3}(n^2 - 13n + 8)$

6 $x(n) = -\frac{11}{15}(-2)^n + \frac{5}{6} - \frac{1}{10}\cos\left(\dfrac{n\pi}{2}\right) - \frac{3}{10}\sin\left(\dfrac{n\pi}{2}\right)$

7 $y(n) = \frac{1}{3} - 6(-\frac{1}{4})^n + \frac{17}{3}(-\frac{1}{2})^n$

8 $f(n) = 5 - 25(\frac{3}{5})^n + 20(2)^{-n}$

9 $y(n) = \cos\left(\dfrac{n\pi}{6}\right) + \sqrt{3}\,\sin\left(\dfrac{n\pi}{6}\right)$

10 $f(n) = (2)^{n-2} + \frac{1}{4}(3)^{n+1}\cos\left(\dfrac{n\pi}{2}\right) + \frac{1}{2}(3)^n\sin\left(\dfrac{n\pi}{2}\right)$

11 $i(n) = -\dfrac{1}{\sqrt{3}}\sinh(\omega_0 n)$ where $\omega_0 = \cosh^{-1}(2)$

12 $x(n) = A + (B - A)n$

13 $i(n) = 2\cosh(\omega_0 n) + \dfrac{6}{\sqrt{5}}\sinh(\omega_0 n)$ where $\omega_0 = \cosh^{-1}(\frac{3}{2})$

14 $f(n) = -\frac{8}{3}(-\frac{1}{2})^n + \frac{8}{3} - 3n$

Alternatively by division of the polynomials $f(n) = 0, 1, -4, -6, -9.5, \ldots$.

Bibliography

Further reading

Doetsch, G. (1971) *Guide to the Applications of the Laplace and z-transforms*, Van Nostrand Reinhold.

Erdélyi, A. (1954) *Tables of Integral Transforms* (Bateman Manuscript Project, 2 vols), McGraw-Hill.

Kaplan, W. (1981) *Advanced Mathematics for Engineers*, Addison-Wesley.

Karni, S. and Byatt, W. J. (1982) *Mathematical Methods in Continuous and Discrete Systems*, Holt, Rinehart and Winston.

Spiegel, M. R. (1965) *Theory and Problems of Laplace Transforms*, Schaum's Outline Series, McGraw-Hill.

Stroud, K. A. (1973) *Laplace Transforms – Programmes and Problems*, Stanley Thornes.

Watson, E. J. (1981) *Laplace Transforms and Applications*, Van Nostrand Reinhold.

Applications

Ahmed, N. and Natarajan, T. (1983) *Discrete-time Signals and Systems*, Reston Publishing.

Aseltine, J. A. (1958) *Transform Method in Linear Systems Analysis*, McGraw-Hill.

Banks, S. P. (1986) *Control System Engineering*, Prentice Hall.

Borrie, J. A. (1986) *Modern Control Systems*, Prentice Hall.

Kuo, B. C. (1975) *Automatic Control Systems*, Prentice Hall.

Schwarzenbach, J. and Gill, K. F. (1984) *System Modelling and Control*, Edward Arnold.

Shinners, S. M. (1964) *Control System Design*, Wiley.

Index